Trees and Forests of the World

Trees and Forests of the World

Why They Matter to Us

MARKKU LARJAVAARA

Associate Professor, Department of Forest Sciences,
University of Helsinki, Finland

Oxford University Press is a department of the University of Oxford.
It furthers the University's objective of excellence in research, scholarship,
and education by publishing worldwide. Oxford is a registered trade mark of
Oxford University Press in the UK and in certain other countries.

Published in the United States of America by Oxford University Press
198 Madison Avenue, New York, NY 10016, United States of America.

© Markku Larjavaara 2026

CIP data is on file at the Library of Congress.

ISBN 9780197757079

ISBN 9780197757062 (hbk.)

DOI: 10.1093/9780197757109.001.0001

The manufacturer's authorized representative in the EU for product safety is
Oxford University Press España S.A. of Parque Empresarial San Fernando de Henares,
Avenida de Castilla, 2 – 28830 Madrid (www.oup.es/en or product.safety@oup.com),
OUP España S.A. also acts as importer into Spain of products made by the manufacturer.

*To my wife and three sons, who I wish could have joined me
on my forest walks around the world.*

Preface

I have experienced the most spiritual moments of my life in forests. For example, I still clearly remember the smells, sounds, and huge, buttressed trees when entering a lowland rainforest for the first time after reading of them for years. This memory is from a six-hectare fragment within Singapore Botanic Gardens, and the experience when revisiting it years later and comparing it to more natural forests was far less spiritual.

Many of us have experienced "wow" moments when staring at majestic trees. Most people like forests and trees, and the readers of this book probably even more so than people on average. Despite this, when pondering why forests have disappeared from a large portion of the globe or how much forestland we want to have in the future, we should aim for an objective and analytical perspective. We should ignore the idea of good guys and bad guys and not consider what our colleagues might think of the chosen theme or its political implications and simply seek the truth. That is what I am trying to do in this book.

Why a book instead of focusing on articles in scientific journals, as most forest scientists are doing? Science is not advancing as before despite increasing resources[1]. Even within disciplines, scientists are clustered into global groups of dozens or hundreds of scientists, with their own premises, objectives, jargon, and sophisticated methods that may be difficult to understand from the outside. Their articles are coauthored, peer-reviewed, and cited by others from the same group, and most do not find time to read publications outside of their own circles. These groups are then able to rapidly advance an area of research but will ultimately reach a dead end, with focus on details and repetitive studies. Large areas between the research circles remain unstudied, and bridges between them are largely missing. More dangerously, some groups suffer from faulty premises, which may be poorly visible from within the paradigm and, if observed, may be quickly forgotten to preserve the harmony from conventional thinking within the research circle. Science is not self-correcting as it should. Therefore, scientists should both write and read more books.

They are more suitable for bringing together various groups than articles are, as articles are read too rapidly to allow time for critical self-reflection.

It is not difficult to figure out why scientists are not writing more books, particularly those with a wider coverage. The systems with how colleagues and employers credit accomplishments are often based on citations by other scientists in the same research circle, or even worse, on the amount of project funding spent. In this book, I focus largely on less studied areas between the research circles and, when touching upon them, I do it mainly with a critical perspective. The motivation for writing this book comes from hope in changing how people think. This is not easily accomplished if faulty ways of reasoning are left unchallenged. My goal is to advance science, and dissidents are valuable for that purpose.

When readers notice a clear divergence from their prior ways of thinking, two typical responses are possible. First, many may defensively seek counterarguments to protect their own ways of thinking. The second common reaction is to follow the novel reasoning without experiencing issues and to even clearly see the divergence from prior thinking but, after a well-slept night, to continue everything as is based solely on that prior thinking.[2] The third and most fruitful reaction would be to simply focus on the facts and to reflect on personal views and compare them to the author's, not only when reading or listening but also after that well-slept night.

When starting my writing journey, I read online advice for first-time authors of nonfiction books. They all advised focusing on a narrow readership. I considered writing only to fellow forest scientists, students, or the well-read public. I pondered this for a long time and ended up writing to all of them at once. I believe this was the right choice, as I mainly focus on the less studied areas in between the well-studied research circles, and I therefore begin by using plain language to explain the basics. The interested public should be able to follow most of the paragraphs but will likely need help by performing some supporting searches online or discussions with artificial intelligence.

The book consists of four parts. They do not cover all forests sciences in a descriptive textbook style with plenty of focus on terminology, but rather zigzag between interesting topics related to the world's forests similarly to an essay collection. In Part I, I begin by describing the world's forests and trees, discuss natural disturbances, and speculate why gymnosperms and angiosperms are distributed as they are. In Part II, I focus on maximal tree size and how it depends on the climate. In Part III, I move from the biological, natural sciences perspective used in the first half of the book to examining how forests benefit humans and how they can be managed to benefit us even more. Finally, Part IV discusses what will happen to the world's forests in the future and what should be done. The second half of Part I and all of Part II are more scientific than the other sections of the book. Readers interested more in why forests matter to people can start reading from Part III onwards and only refer to the first half based on the cross-references in the second half.

Throughout the book, I try to avoid jargon and use simple language. I particularly try to avoid fashionable terms and concepts unless I criticize their usage, such as "climate-smart forestry", "ecosystem services", "forest landscape restoration", and "sustainability", which are often used to vaguely express something positive and occasionally to render analytical critique impossible or to bluff readers into not thinking carefully, especially in policy-related planning and reporting.

For species names I generally use scientific names, written in italics, without the authority. Common names in all languages, and their regional varieties, can rapidly be found online based on the scientific names. In addition to the main text supported with eight figures, I wrote 26 "Sceptic's questions" and 22 "Weird thinking" texts in separate boxes. These texts often do not stay within the boundaries of forest sciences. In the Sceptic's questions, I focus on common scientific reasoning that is potentially flawed. This could had been done more efficiently by using negative citations, but I did not want to go that far. Weird thinking presents novel perspectives that hopefully encourage unconventional future research. I am very direct in them to make my point clear, but please remember that my objective is not to upset people but to quest for the truth.

Acknowledgements

I have been fortunate to have been able to discuss the world's trees and forests and why they matter to us with many teachers, colleagues, and students since I began studying forest sciences in 1995. Most of all, I want to thank Helene C. Muller-Landau, who was my supervisor during my postdoctoral years at the Smithsonian Tropical Research Institute, Panama in 2008–2010. Two years of weekly meetings with Helene led me to understand that it has not always been solely my fault when others have not understood my strange ideas. Also, the project we were managing was extremely interesting, as I was travelling to natural forest plots and instructing dead wood and tree height measurements in Brazil, Colombia, Ecuador, India, Malaysia, Panama, Singapore, Sri Lanka, Taiwan, Thailand, and the United States. The University of Helsinki coordinated another project that allowed me to travel and to realize the many similarities between land-use patterns around the world despite travel writers highlighting their differences, as no one wants to hear about how things are the same. In this project, I computed ecosystem carbon and performed rough biodiversity assessments in human-used landscapes in Indonesia, Laos, Mexico, Peru, Tanzania, and Vietnam. Not all projects were this enlightening for me, and I am extremely thankful to Peking University, which in 2019 finally gave me the academic freedom to focus on topics that I found to be the most important and which I discuss in this book. I began drafting an early and very different version of this book to support my teaching in Beijing, and I am grateful to discussions on topics of this book with undergraduate students at Peking University.

Many colleagues, friends, and relatives commented on earlier versions of this book. I thank them in alphabetical order: Maxime Durand, Cameron Gibson, Jussi Haltia, Toby Jackson, Tapio Lampén, Meri Larjavaara, Tuomas Larjavaara, Tuuli Larjavaara, Tapio Luoma-aho, Olavi Luukkanen, Titta Majasalmi, Jouni Nissinen, "Jake" Payne, Juho Pennanen, F.E. "Jack" Putz, Tanja Pyhäjärvi, Katja Sidoroff, Johan Slätis, and XIE Xiaohan. In addition,

I acknowledge three anonymous peer reviewers, Stella Thompson for linguistic editing, Jodie Keefe and Jeremy Lewis from Oxford University Press, and Hemasankar Arumugam from Integra Software Services. Finally, thanks to Katja, Otso, Aaro, and Veikko for having been supportive in my never-ending quest to understand more about the trees and forests of the world and why they matter to us.

Contents

PART I
THE WORLD'S FORESTS AND TREES

1

Forested Biomes

We humans think that we are significant in the world's ecosystems. This is true when considering how we change them. However, we are not that important when focusing on biomass. All living creatures have abundant element carbon in their structures. Of the total 550 Pg of carbon locked in organisms, only 0.01% is in humans while a substantial 82% is in plants—mostly in woody stems.[1] The massive tree trunks that support leaves high in the sky create remarkable three-dimensional ecosystems, with very different conditions at the tops of the trees compared to ground level. These ecosystems and the biomes that they form are vital for much of Earth's terrestrial biodiversity, to mitigate global climate change, for timber production, and for the livelihoods of over one billion people.[2]

To understand why forested biomes exist we need to understand why trees build their massive trunks, and for this we must examine past evolution. Think of a mighty tree living in an ecosystem that is in a steady state, producing a million seeds during its lifetime. Assuming a perfect steady state and that each large tree is similar, out of the one million seeds only one will become a new mighty tree. Many seeds are eaten or decomposed before they even germinate, and most of those that do germinate die while still tiny seedlings. The remaining compete for light and grow upwards as fast as they can without risking too much. They must build a stem that supports the leaves high up. Luck plays a part, but the offspring are also genetically variable, and the new mighty tree is normally better adapted to the conditions in which they compete than the 999,999 unsuccessful ones on average. The new generation is fitter. Even tiny enhancements from one generation to the next have led to extensive changes during the roughly one million generations that modern trees have had trees as their ancestors.

Trees are extreme because they have benefited from greater heights throughout their evolutionary history. Compare trees to phytoplankton that dominate primary production in the oceans and that control their depth in the water simply by adjusting their buoyancy. Phytoplankton individuals are too small to see with the naked eye, and their populations have complete

Trees and Forests of the World. Markku Larjavaara, Oxford University Press. © Markku Larjavaara (2026).
DOI: 10.1093/9780197757109.003.0001

biomass turnover in just a few days.[3] In contrast, some tree species are over 100 years old when they begin reproducing[4] and may live for thousands of years.[5] Growing large takes time.

The largest trees have not always been so massive. The invasion of land began nearly 500 million years ago with the bryophytes.[6] The evolution of water conducting and the supporting tissues, which enabled greater heights, was relatively rapid, and the first tall terrestrial plants evolved nearly 400 million years ago.[7] Both lycopods and horsetails reached the heights of large modern trees over 300 million years ago. Interestingly, despite similar maximum heights, these two clades and tree ferns, attaining almost similar heights, had completely different tissue structures providing mechanical strength.[8] Early gymnosperms appeared around the time of the tall lycopods and horsetails but gymnosperm trees (hereafter: gymnosperms) began dominating the world's forests less than 300 million years ago.[9] Angiosperms appeared in the fossil records only some 140 million years ago,[10] and eudicot angiosperm trees (hereafter: angiosperms), with a very similar supportive tissue structure,[8] began to dominate certain forested regions around 90 million years ago.[11]

The largest trees are not as massive everywhere, and large areas do not even have trees. Nearly one-third of the Earth's land area is currently forest.[12] Almost a third has been deforested by humans,[13] and the remaining third would be treeless even without humans. These areas are too dry or cold for trees to grow. Sometimes the border between natural treeless land and forest is abrupt, but more typically it is gradual, with small shrubs gradually changing into large shrubs, then into small trees, and finally into large trees when travelling towards a climate that is more conducive for tree growth. In addition to climate, soils also influence whether trees can grow and what size they can attain. Soils and the ways in which trees obtain water and nutrients from the soil vary locally but climates only vary at very large scales, except in mountainous landscapes. Therefore, it is safe to deduce that when a forest changes within a few hundred metres, the alteration is caused by changing soils, unless the human or natural disturbance history is changing as well. For global patterns, climate is the main cause, but the effects of climate and soil are difficult to separate. For example, large areas are so wet with waterlogged soils lacking oxygen that most tree roots cannot grow, and trees are absent or stunted. Such conditions require both high precipitation relative to potential evapotranspiration but also impermeable soils.

Mean annual temperatures are about 27° C in the equatorial tropics at sea level, between 15° S and 15° N, and decrease from there with an increasing pace towards the poles so that a mean annual temperature of 0° C is reached near 60° S and 60° N.[14] Seasonal temperature variation increases with decreasing mean annual temperature and with increasing climate continentality. A one-degree drop in mean annual temperature is, on average, associated with a decrease of one and a half degrees in the temperature of the coldest month and only a half-degree drop in the temperature of the warmest month.[15] Topography also influences temperature. Increasing elevation decreases mean annual temperatures normally between 5°C and 10° C per kilometre upwards.[14]

Air temperatures matter greatly for trees. Plant tissue temperatures follow air temperatures closely,[16] even though water transpiration cools leaves and radiation can both cool them and warm them up.[17] Individual biochemical processes are generally faster the warmer the tissue. Photosynthesis is composed of several processes, including some that consume carbon before carbohydrates are synthesized. The processes may be damaged at very high temperatures. The net result is that photosynthesis peaks at intermediate temperatures. Other processes that consume energy (i.e. respiration), which include material transport, active combat against microbes, and other tissue maintenance, normally increase with temperature. This can be perceived as being bad for the tree, as more energy is spent, but concurrently, the aim of the process causing the respiration is reached faster. Because most other organisms than trees also speed up their actions, the outcome of a defensive situation against a microbe attack may not change, but only the pace at which it is reached is altered (Weird thinking 1.1). Both photosynthesis and respiration acclimatize during the lives of trees and evolve in evolutionary timescales, so that a tropical tree, which has evolved and grown in tropical temperatures, performs better, photosynthesizes more, and respires less in those temperatures than other trees do.

Weird thinking 1.1 How fast does a tree's clock tick?

We humans are endothermic mammals, and our nervous and muscular systems and metabolism are largely independent of air temperature. Therefore, the same clock that we use to measure our performance when

continued

continued

running is also useful for portraying abiotic events such as the length of the diurnal cycle on Earth or the speed of a falling rock. Sometimes we adjust the time, such as to calculate the "human age" of a pet dog based on differences in reaching reproductive age or senescence. However, temperature becomes important when we focus on hibernating or ectothermic animals. For example, a bat hibernating during a winter with scarce food could reduce its metabolic rate by 95%,[48] meaning that if metabolism is used to adjust the timescale, the bat can in a way "shorten" the duration of winter by 95% and a six-month winter energetically lasts just nine days.

Poet Eliot wrote that "The moment of the rose and the moment of the yew tree are of equal duration".[49] Perhaps plant time could be adjusted based on life cycle length, as with dogs. More interestingly, plants can be compared to ectothermic animals, as their photosynthesis, metabolism, and tissue formation are strongly dependent on temperature. Scientists, practical horticulturists and others managing plants have developed various indices, such as heat summation or degree days, in which the temperature difference to a set temperature limit is summed from the days that are warmer than the set limit. These indices are then used to instruct when to apply fertilizers while accounting for the weather of that particular growing season or whether the average heat summation of the area is sufficiently high for a given plant species. Similarly, a widely used global classification of life zones is based on "biotemperature", with temperatures below 0°C and above 30°C excluded and a linear relationship assumed in between.[50] An approach better linking to plant processes, such as one based on the Q_{10} relationship in which a 10°C temperature increase increases metabolisms by a given factor, such as two,[51] could be the basis for calculations and units of time. This would cause changes in results from global analyses of leaf or fine root turnover, and the oldest trees would not be found in the temperate mountains but in the tropics, where the oldest trees have had more potential for metabolism during their lifetime. Similarly, as in the case of the hibernating bat, the leafless season in the boreal would shrink to a tiny proportion of the year, while based on clock time, a tropical dry season would be roughly as long.

Q_{10} time could also reveal why high wood density, which is common for long-lived species, mysteriously increases towards the equator.[52]

It rains much more in the equatorial tropics than elsewhere. Other generalizations about global precipitation are difficult to make. What matters most for plants is soil moisture, which can be estimated based on the difference between evapotranspiration and precipitation, even though soil types and topography also play a role. Trees need some water to photosynthesize, but most of it, taken up by the roots, is transpired while opening the stomata in the leaves to suck in carbon dioxide. Having to transpire can be perceived as good or bad for the tree. Bad because very dry soils prevent trees from photosynthesizing. Good because transpiration is most effective midday when cooling is needed the most, and the more water is transpired the more soluble nutrients are taken up with the water (Sceptic's question 1.1).

Sceptic's question 1.1 Is there too little water for trees?

Textbooks and models normally assume that trees minimize transpiration. Trees need to transpire when photosynthesizing, as some water inevitably escapes when assimilating carbon dioxide through the stomata. Many future simulations predict climate change to reduce soil water and therefore to depress tree growth or cause mortality. These are based on thinking that productivity is boosted by increasing annual precipitation, as indeed is often the case when continental scale patterns are examined[31] or when a drought causes reduced growth or even mortality due to carbon starvation.

However, water is also transpired during the night when carbon dioxide does not need to be assimilated for photosynthesis. The physiological reasons for night transpiration are poorly known, but could be related to nutrients obtained with the water,[32] and the phenomenon is certainly actively controlled by plants,[33] indicating that the simplistic theory of plants maximizing their carbon intake while minimizing water release does not always hold. Second, in many parts of the world, tree biomass and growth decrease with increasing water availability and increasingly

continued

---

continued

---

anaerobic soils. If oxygen is lacking in deeper soil layers, nutrients in these layers are not easily available for trees. For example, five million hectares, or over half of the peatland area, have been drained to reduce soil water and to increase tree growth in Finland.[34] Third, trees often grow well in arid areas that typically only have small trees or no trees at all, if protected from disturbances such as fire and browsing.[35] These disturbances, increasing in severity with declining precipitation, may often be the true reason that biomass decreases alongside precipitation (Chapter 3) instead of being directly caused by lack of water to be transpired. This view is also supported by comparing above- and below-ground biomass investments. Large trees could reach abundant soil moisture or even the groundwater table[36] with a relatively modest investment in deep roots relative to the huge carbon investment in trunks that have to defy gravity and winds.

Sometimes there truly is too little water. However, often it seems as if there is too little but actually there is enough, and trees could grow well if disturbances, such as fire, could be controlled. Quite often there is too much water for trees to grow well.

Tropical rainforests form an iconic forest biome in which plenty of forest remains. Basically, all trees in these forests are angiosperms, and most are evergreen. Typically, only in tropical rainforests do many species develop buttresses at their base when large. Large natural disturbances are rare and, instead, trees die individually, forming gaps of varying sizes depending on how large the dead tree was and whether it pulled down any neighbouring trees possibly entangled in the same lianas, that is, woody climbers. The great variability in the ages and species of tree individuals results in continuous variation in trunk diameters. This differs from the typical boreal forest structure, where one or a few tree cohorts exist with little variation in dimensions. Paradoxically, the complexity of a tropical lowland rainforest, which causes challenges to research regarding individual tree species, enables a rather realistic assumption that the forest is in a stable steady state without temporal variation in tree size distributions or total biomass, but this assumption is not always realistic (Sceptic's question 1.2). Tropical rainforests gradually change into cloud or mountain forests when moving upwards towards colder climates on tropical mountains. Leaves are smaller and more often deciduous towards drier climates with longer dry seasons, buttresses become rare, and species richness and productivity drop (Sceptic's question 1.1).

Sceptic's question 1.2 Are forest ecosystems stable?

Decades ago ecosystems were often faultily assumed to be "in balance with nature" when no human disruption was present.[37] The increasing focus on natural disturbances, such as forest fires,[38] has helped most ecosystem scientists realize how even natural forests are typically recovering from disturbances that occurred decades or centuries before. Unfortunately, the successional status of most forests is not evident to all other scientists, such as atmospheric scientists, who might quantify the carbon exchange of successional forests and attribute the changes to the climate, for example, and not to the varying successional statuses. We forest ecologists are more succession-conscious but still not enough.

Succession in boreal Europe today typically begins after clear-cutting. While clear-cuts and stand-replacing fires both kill trees, the fires create plenty of woody debris and can consume much of the humus layer.[39] Nonwaterlogged soil carbon is widely considered relatively stable due to the near absence of fires. Fluctuations are mainly linked to changes in litter input driven by temporal changes in tree cover, as assumed in Finland's official greenhouse gas reporting. However, assuming a positive correlation between litter input and changes in soil carbon has faced criticism,[40] and relevant to this sceptic's question, omitting the long-term accumulation of soil carbon after a fire may bias estimates of carbon pool changes. In the nineteenth century fires were still frequent,[41] and the initially rapid soil carbon accumulation after a fire[42] may continue for millennia in initially nonwaterlogged soils that either stay aerobic[43] or later become waterlogged.[44] Both waterlogged and nonwaterlogged soils could accumulate carbon similarly, but the waterlogged soils have burned less frequently. Assuming a steady state may be misleading even in equatorial rainforests typically lacking large-scale fires, large windthrows, or insect or pathogen outbreaks. In an active area of research on carbon dynamics,[45] forest ecologists attribute any change in

continued

continued

remote equatorial rainforests to global change drivers such as climate change or increasing carbon dioxide fertilization (Chapter 15). However, even if most natural disturbances remain rare, a drought or flood centuries ago may still influence forest structure and cause it to change. Second, even without disturbances, tree population densities could fluctuate in the absence of human influence. Scheffer and Carpenter[46] stated that even with stable environmental conditions, "fluctuations rather than stable states are obviously the rule". These can be quantified for shorter-lived and more strongly fluctuating populations such as rodents,[47] but longer and less pronounced cycles in tree populations would be impossible to observe after only some decades of surveys and could simply be attributed erroneously to global change. However, these dynamics could also be influenced by humans that may have transported pests and pathogens, such as Dutch elm disease, across continents.

Some forest ecosystems could be stable without human influence, but this is uncertain, and certainly the majority are far from stable.

Large areas of Siberia, Canada, Alaska, and Northern Europe are covered in boreal forests. Most of these forests forming the other iconic biome have burned at least once per millenium.[18] Often succession starts with the sprouting or germination of light seeds of deciduous angiosperms from families Betulaceae, with genera *Betula* and *Alnus*, and Salicaceae, with genera *Populus* and *Salix*. Later, often just one evergreen or deciduous gymnosperm species from family Pinaceae and genera *Larix*, *Picea*, or *Pinus* gain dominance. Draft shrubs from family Ericaceae may form a dense layer at knee height and mosses or lichens a near continuous carpet below them. The northern limit of boreal forests is determined by growing season temperatures and length. The current warming pushes this northern tree line further north.[19] The gradual shift to temperate forests or to treeless steppe in the dry interiors of the Eurasian and North American continents mark the southern or lower edge of boreal forests. Within the area climatically suitable for boreal forest development, waterlogged organic soils with aerobic conditions only at the very surface cause treeless patches of various sizes.

Plenty of forests can be found between the forest-dominated boreal and equatorial biomes. However, large areas have been deforested for agriculture, some several millennia ago like the plains of eastern China,[20] and we do not even know what kind of forests would grow there naturally. Croplands and pastures increase markedly both at the warmer perimeter of boreal forests and the dry edge of tropical rainforests. The remaining forests, often found on slopes that have been more difficult to deforest, are functionally, structurally, and taxonomically intermediate between boreal forests and tropical rainforests.

Generally, climates and therefore vegetation are similar in equivalent latitudes in the Northern and Southern Hemispheres. However, large land masses are missing from the southern latitudes that lie at a similar distance from the equator as where boreal forests are found.

Imagine that a city child, who has watched dozens of nature documentaries and paid some attention to the vegetation in which lions and bears roam, is brought blindfolded to a forest somewhere on Earth. Does she know where she is when the blindfold is removed? If she is in a boreal forest or a tropical rainforest, she would guess the biome right, but knowing the correct continent would be challenging. How much better would I be with all the experience that I have? I can identify most boreal gymnosperms and quite many angiosperms and I roughly know their distributions, so unless I am dropped in a plantation with exotic species, I would know the continent and the side of it. Tropical rainforests would be trickier, as I cannot identify almost any of the tree species, and I might not even get the continent right. Looking up and spotting tall trees with narrow crowns would make me bet on southeast Asia,[21] and looking down and recognizing the large, winged seeds from a tree belonging to the family Dipterocarpaceae would lock in my guess. Seeing a palm would suggest that I am in a neotropical (New World tropical) forest.[22]

I have been organizing a hand-on exercise of the world's forests on a lawn for secondary school pupils visiting our university. Before they arrive, I have demarcated the roughly equal areas of boreal, temperate, and tropical forests, so that a square metre of the lawn corresponds to one million square kilometres in the real world. I have also collected a pile of 276 leaves in the middle. I have then asked the teenagers to divide the leaves into three piles corresponding to the number of tree species in each forest zone. All groups have known that there are more species in the tropics but none how extreme the imbalance is. Despite the three zones being about the same size, one leaf should go into the boreal representing 161 species, eight into the temperate zone, and the remaining 267 leaves into the tropical representing a whopping 43,000 tree species.[23] A more recent calculation, based on a very broad definition for trees where only a height of 2 m was required for single-stemmed woody plants, amounted to 73,000 species, 31,000 of which are South American.[24] This estimate included as-of-yet undiscovered species.

The drastic decrease in the number of tree species from the equator to the poles has been a central theme, possibly even too central, in forest ecological and biogeographical research. Often the question has been why tropical rainforests harbour so many species, revealing that most botanists have grown up in the north and therefore consider the low northern species diversity normal and not needing an explanation. Diversity would increase over time with mechanisms that benefit rarer species, as new species evolving or migrating to the landscape is initially rare, and established species would not go extinct as easily because their competitive advantage increases before extinction, unless the collapse is sudden. Pests and pathogens that attack only a portion of all tree species cause such a mechanism if the likelihood of spread from one host tree to another decreases with increasing distance. To support this, seedling survival has been shown to increase with increasing distance from the parent tree.[25] But such patterns are similar in all biomes,[26] which does not help in explaining why so many fewer species occur far from the equator.

Dozens of theories have been developed to explain the latitudinal gradients.[27] According to one particularly convincing theory, the relative instability of climatic conditions away from the tropics has caused more extinctions and therefore reduced the current number of species. However, it would be unrealistic to argue that tree species richness would, assuming the same past climatic stability, remain constant right to the treelines delineated by coldness or dryness. The mechanisms causing the latitudinal gradient described in the theories[27] are normally not mutually exclusive and most may be significant for some clades and areas. For example, animal seed dispersal and mycorrhizal association may play a role.[28] Size may also matter, as smaller plants have shorter generations speeding up evolution and speciation,[29] potentially decreasing the number of large plants relative to small plant species at high latitudes. This prediction is strongly supported when focusing on the proportion of woody species, which are larger and represent about half of all species in tropical rainforests but only about one-tenth in the boreal biome.[30] Size may also influence how trees cope with the bad season (Weird thinking 1.2). There are also many additional mechanisms at play that influence species groups and their relative biomass contributions differently, thus indirectly impacting species richness (Weird thinking 1.3).

Weird thinking 1.2 Does size help survival during the bad season?

In many seasonally cold regions of the world families celebrate the start of the summer season by dressing up in light clothes and going out to eat ice cream. Often this is a head start, and the temperature would be more suitable for drinking hot chocolate in a thick jacket. The kids craving their ice creams initially look excited, but as the eating continues they begin looking progressively like they are freezing while their parents look more or less the same from the start till the last gulp. Relative to the adults, the kids melted and warmed many times more ice cream relative to their body masses. Kids need extra energy for their growth and, relative to other foods, are likely to like ice cream more than their parents do. However, interestingly even full-grown smaller mammals need more energy per unit of their mass to simply maintain a given size. The metabolic rate scales approximately to body mass to the power of two-thirds,[53] meaning that doubling the mass increases energy consumption only by 59%. Because energy reserves depend linearly on body mass, larger mammals can endure longer with their own energy stores. The huge Kodiak bears during the long off-season when salmon are not easy to catch in rivers lose a smaller proportion of their biomass than small mammals with a similar seasonal cycle in their diets do. Larger mammals can also have a thicker fur without it greatly hindering their movement, which could be an additional factor that boosts size in seasonally cold climates.

Larger trees need more energy to maintain themselves but not as much as expected based on their mass. Tree metabolism scales roughly to a mass to the power of three-quarters.[54] This means that doubling the mass would result in 68% higher energy consumption. If the size of the reserves that trees have for the bad season depends on tree size, larger trees could survive the winter or dry season more easily just as the enormous Kodiak bears do. However, trees and mammals are different from each other. Animals need to move and a larger than necessary body

continued

---

continued

---

volume is unpractical, and hence, all mammals have body densities close to the density of water. In contrast, trees have variable densities of wood, and most trunks contain plenty of unused space. Cell walls in wood are 50% denser than water,[55] but even fresh wood normally floats,[56] revealing that living trunks have gases in them. As trees do not need their whole trunks for storage, the greater ratio of volume or mass to energy consumption should not help trees to survive the bad season. However, larger trees may be able to grow deeper roots than smaller plants can, and this could help them survive a long dry season (Chapter 12). On the other hand, small plants have other means to survive in seasonal climates (Weird thinking 1.3).

Weird thinking 1.3 Are there other ways to cope with seasonality?

Even if the climate is optimal for one month a year, at other times the temperatures can be too low or high or precipitation can be insufficient or cause flooding. Animals have several options to mitigate the energetic penalty caused by unfavourable seasons. They can migrate to regions with more favourable climates, such as migratory birds, hibernate like northern bears, or complete their full life cycle in good weather and persist through the bad seasons as larvae or pupae such as most boreal insects. Smaller animals reach maturity when younger,[57] and undergoing the entire life cycle within a season is possible for flies but not elephants.

Plants are always similar to hibernating animals in that they do not move and their tissue temperature closely matches that of the ambient air. Small plants can complete their life cycle within a favourable season and persist as seeds or underground tubers, just as small animals do as eggs, larvae, or pupae. Because nearly all tree species need decades to reproduce efficiently, they are like large mammals that cannot benefit from surviving the winter or dry season as seeds or tubers and rapidly bounce back during the good months.

Even if the differences in how various species groups cope with seasonality are small, they can still have drastic impacts on relative species numbers or biomasses if the groups are ecologically similar, and a small competitive advantage can help a species dominate in the ecosystem. The number of bird species can be so numerous during the boreal summer relative to other vertebrates in boreal forests[58] because they are able to migrate south for the winter. Small, annual herbaceous plants suffer less from seasonality and should therefore compete better with woody plants in a seasonal climate than in an aseasonal one, even if the climates are similar, on average, from an energetic perspective. This could increase the need for weeding after clear-cutting and planting in practical forestry in more seasonal climates.

Speculations of how trees might interact with animals are even more challenging to make. When focusing on trees, their insect herbivores, and on the bird predators of those herbivores, the herbivores could be the winners in a very seasonal climate, as the predators cannot complete their life cycles within a growing season and must instead migrate south or struggle to survive the winter on-site. This could partly explain large-scale insect outbreaks in the boreal forests. However, the story is not so simple, as trees can defend themselves to mitigate herbivory by increasing defensive chemicals, which I therefore expect based on this oversimplified view to increase with seasonality. As rare species suffer less from species-specific herbivores, increasing the importance of herbivory should increase tree species richness, but this goes against the observed global gradients.[26] However, seasonality reduces species richness via other mechanisms,[27] as I explain in the last paragraphs of the main text of this chapter.

2

A Multitude of Trees

Is a banana tree actually a tree? Normally the answer is no; rather, it is a giant herb. My answer is that it depends on the definition of a tree. The definitions are language- and culture-specific, but they are commonly based on the internal structure of the stem, the size and form of the whole plant, or both simultaneously. Rarely do people refer to evolutionary or taxonomic relatedness when claiming that a given plant or species is or is not a tree. I do not argue for the superiority of one definition over others, and I do not follow a particular definition in this book. Nevertheless, to support discussions beyond this book, it is useful to discuss the dimensions that influence whether we consider a given plant to be a tree or not.

Lignin, or plant tissue woodiness, is often a requirement of the stems, that is, the trunks of trees. This lies behind the logic of not considering the banana plant to be a tree, as its stem is herbaceous, that is, lacking lignin. Other monocots, such as palms (family Aracaceae) and bamboos (subfamily Bambusoideae in the family Poaceae of grasses), are typically not considered trees despite their stems containing lignin. Their exclusion is not taxonomically justified, as they are more closely related to clade eudicots (eudicot trees are called "angiosperms" in this book even though monocots are angiosperms as well), to which most trees belong, than to gymnosperms, which unquestionably include trees. However, palms and bamboos have stem growth patterns that differ drastically from those of typical trees. "Normal" tree growth involves trunk thickening, branch formation, and expansion at the tips of the stem and branches. Most palms cannot thicken their stems, and most species have single stems without branching.[1] Their stems are not composed of wood typical to trees; rather, the densest tissue is found closer to the surface to increase strength. Bamboos differ even more from typical trees because their stems are normally hollow. Furthermore, bamboos increase their height not only from their tips, because their stem sections, that is, internodes, also expand. Whether it is a good idea to classify bamboos as trees depends on the context. For sawmills, bamboos are not trees, but their hollowness does not matter for hydrological modelling,

Trees and Forests of the World. Markku Larjavaara, Oxford University Press. © Markku Larjavaara (2026).
DOI: 10.1093/9780197757109.003.0002

which instead cares about the treelike three-dimensional structure and leaf area. Therefore, from a hydrological perspective, bamboo thickets are often better clumped into the class of forests.

Another commonly perceived borderline between trees and nontree plants is related to size and overall form. The seedlings of woody plants that most people agree are "trees" when large are normally considered trees as well, even when small in size and possibly still herbaceous. Some species have individuals that develop into large plants that most people clearly consider to be trees, while in other conditions they are regarded as lianas or shrubs.[2] Shrubs are typically defined as woody plants that never develop a single trunk that supports leaves high up. Curiously, it seems that woody plant heights do not range evenly from draft shrubs to canopy trees, but rather plants around eight metres tall are rarer.[3] The division between trees and shrubs is similar in many languages and does not seem to be biome dependent. This height distribution pattern can be understood based on biomechanics, as shrubs can be bent all the way to the ground and are therefore protected from the strongest gusts because they can bounce back unharmed.[4] Being tall and being able to bend to the ground is biomechanically impossible; therefore, it is normally bene-ficial for trees to build a single massive trunk, thanks to which they can resist with brute force the strongest wind gusts. People can climb trees, but shrubs bend to the ground under a person's weight. A shrub's biome-chanical strategy does not allow for reaching a decent height with much branch and leaf weight, as adding stem diameter to support these struc-tures would prevent shrubs from bending to the ground without breaking. Instead, shrubs need to grow another stem to add leaf area. Multistemness is so typical for shrubs that it is included in many definitions. On the contrary, trees benefit from focusing on adding biomass to a single stem (Sceptic's question 7.2).

The definitions of a tree and a forest go hand in hand. An ecosystem dom-inated by trees is normally considered a forest. However, some distinctions must be drawn between a forest and nonforest. Open savannahs in dry cli-mates may or may not be considered forests, similarly to sparse stands close to the northern treeline, even though both have trees. The main plantation products may influence whether a stand is considered a forest. Despite there being no disagreement that apple trees are trees, a dense apple orchard may not be a forest to most people, as its objective is to produce food and not wood. Similarly, rubber plantations from tall *Hevea brasiliensis* trees, with no aim for timber production, may not be considered forests. Finally, only a few would call a small isolated group of trees a forest, and for such patches to be accounted in forest area the minimum size requirement has varied from 0.05 to 1.0 hectares.[5]

More interesting than splitting hairs over definitions is to contemplate why trees are as they are. The evergreen or deciduous nature of tree species is often used to describe forests of a given biome (Chapter 1). To move from this descriptive level to the mechanisms that explain whether ever-greens or deciduous species are better at competing than the other group is, we need to first consider the optimal leaf lifespan in an aseasonal cli-mate with favourable conditions for photosynthesis every day of the year. What matters next are the construction costs of a new leaf, how its pro-ductivity drops as it ages, and how the tree grows, influencing how old leaves shade new leaves and require support from the trunk. In fertile soils, the optimal lifespan is short, as acquiring necessary nutrients is efficient[6] (Weird thinking 2.1). In addition, growth speed may influence leaf lifes-pan because the leaves of fast-growing trees may end up in shady locations faster and are therefore shed sooner. Leaf herbivory may reduce evergreen fitness, as the continuous food resource evergreens offer may uphold high herbivore populations. Even with a continuously favourable climate, it might be better to shed the leaves and grow new ones soon after to decrease herbivore populations by interrupting their otherwise continuous food source.[7]

With seasonal variation, understanding whether evergreen or decidu-ous tree species have a comparative advantage becomes more complicated.[8] Cold or dryness can halt photosynthesis or decrease it so that maintain-ing the leaf consumes more energy than it produces. Leaf shedding may still not be advantageous if the energy saved cannot compensate for the construction costs needed for the new leaves. Most tree species in the trop-ical rainforests and cloud forests with very little seasonal climate variation are evergreen. Dry tropics with a clear dry season are often dominated by deciduous species. In temperate and boreal forests the patterns are more complicated, with gymnosperms being typically evergreen and angiosperms deciduous. Evergreens are common in Mediterranean climates with dry summers, probably because the spring and autumn growing seasons are short and flushing leaves twice annually would be costly. These patterns are influenced by human preference for evergreens in plantation forestry (Sceptic's question 2.1).

Sceptic's question 2.1 Why do evergreens grow faster?

Forested biomes, except for the equatorial tropics, have a significant proportion of both evergreen and deciduous tree species. We lack global
statistics on whether evergreens are more common in plantation forestry
than deciduous trees are, but this is apparently so and has been attributed
to the faster growth rate of evergreens. A textbook explanation for the
faster growth rate of evergreens is that they benefit longer from the
investment made to their leaves. Alternatively, from a more ecosystemic
perspective, evergreens photosynthesize for longer, for the entire year in
certain climates, and therefore produce more. Other things being equal,
these explanations hold, but other things are not equal; instead, the world
is a chaos of never-ending chains of impacts. For example, leaves may
produce less than they consume and can be damaged, or support pests
and pathogens during the unfavourable season. Furthermore, building
a long-lived leaf is more costly because more physical and chemical
defences are needed.

The chaotic worldview becomes clearer when we remember how trees
have evolved competing for light by applying height growth (Chapter 7).
In a given area, deciduous species have competed successfully with
evergreens. Otherwise, they would have faced local extinction. How is
it then possible that foresters value evergreens for their rapid growth
if all growth is similar from an evolutionary perspective? Forester's
growth is measured in trunk volume or biomass, while biologically
trunks are needed to safely support the crown at an elevation. If the
crown is leafless in winter when a trunk needs to resist not only the
usual wind friction and the tree's own weight but also the weight of
snow and ice attached to its branches, a thinner trunk is sufficient for
a given growing season's tree height and crown size. Therefore, deciduous trees compete for height equally well with weaker and less valuable
trunks.

To study how evergreenness has influenced species selection in plantations, the importance of winter snow and ice could be compared
regionally to plantation species statistics. I would expect the popularity
of evergreen species to be highest in climates with abundant snow and
evergreens, and deciduous species to be used in similar proportions

continued

continued

as in a natural forest tropical climates. However, as gymnosperms and angiosperms are very different in numerous ways (Chapter 4), the comparisons are best done only for angiosperms.

Leaf size is another tree trait that is easy to observe, and a key axis of global variation in tree form (Sceptic's question 2.2). It increases with increasing temperature and precipitation, being highest in tropical rainforests.[9] Explanations for this global variation have been based on micrometeorological theories. The boundary layer, with minimal air movement very close to the leaf surface, is thicker far from the leaf edge, thereby reducing its exchange of thermal energy with the surrounding air. The middle parts of large and round leaves therefore heat up easily in the sun and correspondingly cool down during cloudless nights.[9] In addition, the thick boundary layer reduces transpiration. Temperatures in all parts of small leaves or of larger but lobed leaves close to the edge fluctuate less diurnally. This can be particularly valuable in cold climates to avoid leaf freezing during growing season nights.[9]

Sceptic's question 2.2 Should one large leaf be compared to one or two small ones?

When someone is pondering whether to buy a small rather than a large car, it is easy to list the pros and cons to be compared. Small cars are cheaper to purchase and use and are more environmentally friendly but less spacious, and give less of a status boost for the driver. Similarly, small leaves are energetically cheaper to build and cause less wind friction, but can photosynthesize less. However, this type of comparison can be misleading.

The car example makes sense, as people normally buy one car at a time. However, when considering optimal leaf size, the number of leaves per plant is not fixed. With this misleading way of thinking, it is easy to conclude that larger leaves are better for photosynthesizing rapidly, but need a stronger trunk to resist the force of the wind and abundant resources to be used in growing and maintenance. Instead, when fixing the point of comparison more fruitfully to a given combined leaf area, the comparison becomes more relevant. Assuming that thickness is unrelated to leaf size, a leaf twice the size is about twice as costly to build and photosynthesizes about twice the rate, but the focus should be on the subtle deviation from the "twice", which is the basis for understanding the global variation in leaf size.[9] For example, the photosynthesis of a single large leaf may decrease during the middle of a hot day due to higher temperatures in its central parts.

Instead of fixing leaf area, comparing small and large leaves with the same construction or maintenance cost for a tree would be an even better method, as those are processes that have been directly relevant in the evolutionary history of trees. However, they are challenging to quantify, and leaf area is hopefully a satisfactory proxy for construction and maintenance costs. Leaf biomechanics are complicated,[20] but should be tackled before we can be more confident of our explanations regarding global leaf variation.

The micrometeorological approach to comprehending leaf size variation does not provide the full picture. First, we lack understanding of leaf size from a biomechanical perspective.[10] It seems unlikely that when assuming a constant leaf tissue temperature, the leaves of the world's trees would be of equal size. Instead, it is likely that the relative costs of construction and maintenance (Weird thinking 2.1) influence what is biomechanically the best design for the thin twigs supporting the leaves. Similarly, a tree's overall crown architecture (Chapter 5) is likely to influence its optimal leaf size, but we are far from understanding how.

Weird thinking 2.1 How do trees cope with nutrient scarcity?

Most animals get all they need via their mouths, but trees obtain water and nutrients via their roots and their leaves use light to photosynthesize carbohydrates composed of carbon, hydrogen, and oxygen, which can be used for energy or to build structures. When water or nutrients are scarce, more roots are needed relative to the above-ground parts. Individual trees can acclimate and grow more roots when needed, but larger

continued

continued

roots evolve slowly, with species being adapted to dry or nutrient-poor conditions.[21] However, trees have another completely different way of coping with nutrient shortages.

A tree leaf is first constructed and then maintained until its senescence. The construction requires carbohydrates but also other elements obtained as nutrients. However, maintenance mainly only requires energy obtained from carbohydrates originating from photosynthesis. A long-lived organ requires less maintenance relative to construction and therefore less nutrients are needed relative to carbohydrates or energy. The lower the soil fertility, the less trees can draw nutrients, and therefore it is better to build long-lived structures. Most of the toxins that plants use to protect themselves from herbivores and to make themselves long-lived contain these rarer elements obtained as soil nutrients. Therefore, paradoxically, with less of these elements in the soil, there are more of them in tissue as defensive compounds.[22] However, these same elements are also needed for other functions such as photosynthesis.

Soil fertility influences not only the turnover of certain tree parts, such as the leaves, but also the turnover in the whole ecosystem, with impacts to maximum tree ages and biomasses. Therefore, despite higher soil fertility, such as due to fertilization or increases in the maximum individual sizes of a given species, lower fertility can result in slower life history strategies (Chapter 3) and slower overall ecosystem turnover, and these can lead to larger biomasses. This is the case in the Amazon biome, where greatest biomasses[23] are found in the low-fertility[24] eastern part with many tree traits, such as high wood density[25], associated with a slow life history strategy.

The second major issue not commonly discussed when considering optimal leaf size is the shading effect. The full shade of a large round leaf extends much further down, and this is problematic for the leaves beneath because photosynthesis nonlinearly depends on photosynthetically active radiation, or light. For example, having two leaves in 50% full light is photosynthetically much better than having one leaf in full light and the other in darkness. This mechanism is not important on a cloudy or hazy day, when diffuse light dominates, but is significant in sunny conditions. For a given amount of light, *Pinus sylvestris*, with tiny needle-like leaves, loses less of its GPP (gross primary productivity) when the day is sunny than a broad-leaved species loses.[11] For this same reason, the greatest leaf area per unit land area is measured in forests with tiny leaves.[12]

The reduced photosynthesis beneath the crown of large leaves is naturally a problem for the same individual if it has more leaves in the understorey. If not, then the weaker light underneath is only positive for the larger tree, as it reduces the below-ground competition for water and nutrients and lowers the likelihood of small tree individuals pushing through the canopy in the future. Based on this reasoning, large-leaved species would be expected to have wider crowns, as they are able to block their neighbours even with modest shade. However, not all leaves are horizontal or of the same size. Typically, the top leaves are smaller and more vertically positioned, both of which reduce the risk of full shade beneath, but the verticality increases the level of shade towards the sides with neighbouring trees.

In addition to leaf characteristics and the overall branching architecture (Chapter 5), bark colour and other features are useful for identifying tree species and make us wonder why these differences exist. Bark serves many roles but principally exists to protect the inner tissue from pathogens, herbivores, and forest fires.[13] Bark thickness is largely determined by the fire regime, as increasing thickness increases the thermal insulation and therefore protects trees from flames. To adapt to surviving surface fires, trees have thick bark only at their bases. To survive more intensive fires with higher flames, upper parts also have thick bark. However, trees adapted to extreme crown fires have thin bark because they die in a fire regardless of bark thickness.[14] In addition to these roles that protect the wood, slippery or flaking bark can effectively protect the whole tree from a liana invasion. Bark also contributes mechanically and significantly stiffens thin branches. However, this is clearly only a secondary function, as it is weaker than wood[15] for a given biomass investment.

Surprisingly little has been written of bark colour despite it being very visible and potentially important in tree evolution. Most species in a tropical rainforest have brown bark, providing only a little help to people facing the often too demanding task of species identification. On the contrary, boreal trees have highly variable bark colouring, making the easy identification even easier. Particularly striking is the very light colouration of several species in genera *Betula* and *Populus*, which has been explained to be an adaptation that makes dark insects potentially harming the tree more visible to their predators[16], but none of the explanations clarify why this only occurs at high latitudes. If heating darker surfaces in sunlight was an issue, tropical species could be expected to have light-coloured bark to avoid the extra maintenance costs due to the wood warming up, yet this is contrary to observations. Stem photosynthesis, which is important in the species-rich genus *Eucalyptus*[17], may significantly influence the colouration.

The discussion regarding tree age is less straightforward than initially thought. Sprouting from the trunk base is common, whereas it is less typical from the roots, yet significant in the genus *Populus*, for example. Branches buried in the litter can develop into new trunks in the genus *Picea*. Are the trunks developing from sprouts, root suckers, or buried in litter just expansions of the same individual, or should this be considered asexual reproduction (Weird thinking 3.1)? When the focus is narrowly on the maximum ages of individual trunks, over 2,000 growth rings have been counted from the trunks of seven gymnosperm species from families Cupressaceae and Pinaceae.[18] In contrast, not even 1,000 rings have been counted for angiosperm trunks as their heartwood decays easier, but more unreliable radiocarbon dating suggests that some may reach similar ages as gymnosperms do.[18] The oldest gymnosperms grow in many climates[18], and the global variation in the age of the oldest trees in forests are surprisingly small. Exceptionally old trunks in the Amazon rainforest may reach 1,000 years of age, but even reaching 400 years is already very old.[19] Typical senescence and degradation of the fastest species reaching the main canopy height is somewhere around 50 years in lowland tropical forests, which is similar in the boreal even though fast species typically only occur in early succession when the main canopy is lower. Correspondingly, the very oldest boreal trunks may reach 1,000 years of age. These ages lead to extremely slow turnover rates compared to shrublands, grasslands, and especially to aquatic open water ecosystems (Chapter 1).

3

Natural Disturbances

Natural disturbances are normally thought to include unusual natural events of reasonable size, lasting a limited time, and causing mortality. Much of the natural disturbance literature focuses on fire and windstorms. Other often discussed disturbance agents are landslides, floods, droughts, pathogens, pests, and herbivory by vertebrates. All the aspects of the definition are continuums, and the exact definitions of natural, reasonable size, limited time, and mortality depend on the case. For example, human-caused uncontrolled fires are normally considered natural disturbances but prescribed burning is not. A single tree dying may not be considered to be caused by a natural disturbance but a group of 10 large ones might already be. Even if the focus is on forests and trees, a fire not killing trees might be included in discussions regarding natural disturbance dynamics, as it significantly impacts other plants and soil. However, windstorms, for example, that do not cause mortality would not impact those factors and would not be defined as a disturbance. Finally, the limited nature of the time period can also be discussed. Fires and windstorms are normally over during the same day, but floods can last for weeks and droughts for years. However, a "permanent drought", such as in an arid climate, is rarely if ever considered a natural disturbance.

Large natural disturbances that kill most large trees are very rare in tropical lowland rainforests. Expectedly, local catastrophic events, such as volcanic eruptions and landslides on steep terrain, can cause total ecosystem destruction, but these are very rare in all biomes. Unusually wet years can cause floods in low-lying landscapes, but only tree species able to tolerate anaerobic soil conditions will remain in these areas if such years are frequent. These conditions do not therefore cause dramatic mortality. Droughts kill some trees and have been shown to widely impact the Amazon basin by reducing the mean canopy height and therefore the biomass.[1] Litter and understorey vegetation have relatively high moisture contents even after a long drought due to relatively high air humidity.[2] However, fires have possibly killed large areas of tropical lowland rainforest trees around once a millennium after months of extreme drought. Windthrows can be found in

Trees and Forests of the World. Markku Larjavaara, Oxford University Press. © Markku Larjavaara (2026).
DOI: 10.1093/9780197757109.003.0003

all forested biomes, causing both uprooting and trunk breakage. However, even windthrows are characteristically small, often consisting of only one mature tree.[3] Sometimes a large falling tree that has been connected by lianas to neighbouring trees will pull those down as well. Larger gaps can occur due to severe downbursts associated with thunderstorms.[4] Even larger openings can be caused by tropical cyclones, but these do not occur close to the equator and are common at a 10 to 30 degrees latitude from it.[5] Most tropical lowland rainforests are very rich in tree species, and the gaps caused by most pests and pathogens, specific to one or a larger minority of all species, are typically very small.

While precipitation seasonality has little effect on fuel loads, which are more closely tied to total annual precipitation, the amount of rainfall during the driest season largely determines how flammable those fuels are during the peak fire season. Farther from the equator, ecosystems still receive abundant annual precipitation that supports high biomass and fuel loads; however, as the dry season grows longer and more arid, fires become increasingly common.[6] As local species have evolved to withstand frequent fires, occurring almost annually in some locations, large trees can normally survive the fires well. However, seedlings and sprouts of even fire-adapted species are vulnerable to fires, and species have variable means of coping with this. Despite the adaptations, the above-ground parts of small trees and smaller vegetation often die off and only large trees survive, creating a size structure typical of savannahs, for example. Then, large tree renewal may occur in cohorts during exceptionally long periods without fires or at least without severe fires. It is also important to understand that most ignitions in most ecosystems are human-caused, and fire regimes solely from natural ignitions could therefore be completely different. Many Australian temperate forest ecosystems experience stand-replacing fires, that is, fires that kill most of the large trees. However, forests in similar climates on other continents may experience frequent surface fires, which is common in the dry tropics. Such ecosystems are typical in the western United States, with *Pinus ponderosa*-dominated sparse forests. In some temperate regions, fires are naturally nearly absent, such as in parts of the eastern United States and Central Europe.

Much of the discussion concerning natural disturbances focuses on boreal forests where, despite the high relative humidity during most of the year, forest fires and other very large-scale natural disturbances can be very influential. However, a clear distinction occurs between the North

American and Eurasian boreal. The North American boreal is typically characterized by large stand-replacing fires and, curiously, also large insect outbreaks and mortality due to pathogens.[7] Interestingly, the Eurasian boreal experiences much fewer stand-replacing fires.[8] The Eurasian low-intensity fires are typically carried by mosses, lichens, and litter but also consume dead wood and shrubs, while leaving especially large trees of genera *Larix* and *Pinus* alive and nearly unharmed. Intriguingly, other large-scale natural disturbances also seem to be much rarer in boreal Eurasia.

The above discussion on natural disturbances and how they influence forests and trees has remained at a superficial level. More interestingly, as trees have evolved to cope with natural disturbances, the less discussed perspective is focusing on the exclusion of unsuitable species, the evolutionary adaptation of a given species, and the acclimation of individual trees during their lives. Outer bark thickness is often presented as an adaption to slow down the heating of the inner bark and hence to enhance fire survival. Paradoxically, a thick-barked tree may actually benefit from a fire if its neighbours have thinner bark. Similarly, an herb may benefit from being grazed by sheep if surrounding plants suffer more extensive damage (Weird thinking 10.2). In addition to a thick bark, secondary compounds in leaves and other tissues are frequently explained to be toxic to potential herbivores. However, unfortunately the logic regarding wind mortality is, typically, that trees are what they are and exceptionally strong winds topple them over, without consideration given to wind acclimation.

Disturbance agents vary drastically in whether a tree can acclimate towards them during its lifetime. Fires and wind represent extremes, while biotic agents, such as mammal herbivores or pathogenic microbes, are somewhere in between. In most forest ecosystems, the mean fire interval is much longer than the lifetime of an individual tree, and acclimation based on past fires is therefore typically impossible. Wind is different, as there is wind nearly every day and trees build a thicker stem if exposed to strong winds.[9] This has important implications for understanding climate change–induced wind mortality risks (Sceptic's question 3.1). Biotic disturbance agents are often intermediate, although some lethal ones may appear only once during the life cycle of old trees but others may be continuous or annual, thereby allowing trees to acclimate to their presence (Sceptic's question 15.2). In addition to the acclimation, other mechanisms may also be mitigating the initial shock. It may take time for the predators or pathogens of the disturbance agents to increase in numbers.

Sceptic's question 3.1 Is the windthrow risk increasing with increasing top wind speeds?

Most scientists in the field—and there are a lot of them[16]—would say that of course it is. However, I argue that it depends on the situation. Trees acclimate relatively rapidly to wind changes.[17] Therefore, if the acclimation is complete, so that windthrow trees adjust their trunk diameter, height, and crown sail area to the new conditions, the windthrow risk remains unchanged. Full acclimation takes time, at least months but possibly years, and does not happen during an exceptionally stormy day, which therefore often causes wind damage. Similarly, the suddenly altered winds after logging increase the windthrow risk, but after some time the trees have acclimated to the new conditions and the risk is back to normal.

Even with plenty of time for acclimating, the windthrow risk may be increased. For example, if maximal winds caused by tropical cyclones increase with climate change, while normal strong winds do not, trees cannot acclimate to the rare events, in which case much more mortality will ensue than caused by the previous, weaker tropical cyclones. Similarly, exotic species can be vulnerable to strong winds, such as *Picea sitchensis* in the British Isles,[18] possibly due to stronger peak winds relative to normal winds that trees use to acclimate their structures. Trees could potentially be unable to acclimate based on wind swaying during their dormant season, and increases in dormant season winds could prove fatal. However, these mechanisms are highly uncertain and could be studied with recently developed methods.[19]

Generally, simply assuming that a stronger wind means more wind-caused mortality, as is assumed in dozens of scientific articles published annually, is misleading. Instead, in conditions that have remained relatively stable for at least a couple of years, it is better to model the wind mortality risk based on the competitive status of the individual tree. Trees that have a greater risk of being outcompeted by their neighbours in the upward race towards the light take a greater risk of being windthrown. This is the reason why occasionally, when a forest patch in the middle of an open area is hit by strong wind, all the trees in the centre, which compete vigorously for light, are snapped or uprooted, while trees in the perimeter survive despite the stronger winds, as they have acclimated to them and compete less for light. Similarly, wind speeds do not explain which trees snap or uproot, but instead, the faster species with lower wood density are the ones that are killed by strong winds,[19] and not only by snapping but by uprooting as well.[20] The uprooting demonstrates that low wood density does not directly cause snapping, but is associated with a fast life cycle and high risk.

Trees have an extremely long life cycle compared to phytoplankton in the oceans (Chapter 1). Even among terrestrial vascular plants, trees are extremely slow. Annual herbaceous plants complete their full life cycle within a year and other herbs and shrubs possibly within a few years, but trees need more time on average. Large differences occur between tree species; some are closer to other plants and begin reproducing young and die young, while other species are slow. Many traits are associated with a slow life cycle, such as long leaf lifespan[10] and high wood density.[11] The case with sexual and asexual reproduction is more complicated (Weird thinking 3.1). Placing tree species along the fast–slow continuum has been the basis of much forest ecological research.[12] Fast species are sooner able to utilize post-disturbance light and nutrients and to thrive early in succession. On the contrary, slow species are unable to rapidly exploit surplus resources but grow slowly, have low mortality, and can compete well with low resources. They have low mortality because they invest more in resisting both continuous mortality agents, such as decay-causing pathogens, and disturbance agents. In human terms, slow species are "far-sighted", and a theory could be developed to describe the far-sightedness of plants using interest rates (Weird thinking 3.2). When the tree is slow, its evergreen leaves and other parts also have long lifespans.[10] However, not all tree species fall nicely along the fast–slow continuum, as some have both high mortality and slow growth but do not go extinct thanks to their intense reproduction.[13]

Weird thinking 3.1 Sexual or asexual reproduction?

Why do organisms need to reproduce? Obviously, population growth would not be possible without reproduction; however, even in stable populations, the replacement of deceased individuals is necessary. If you

continued

continued

---

are over 50 years old, it is not difficult to imagine how poor an idea eternal life would be. Consider the ailments your body already suffers from, and then imagine all those that would accumulate over several millennia of active life.

Reproduction clearly is necessary, but why does sex exist?[21] Sexual reproduction is complicated yet common. Genetic mutations drive evolution. However, most mutations either have no effect on fitness or are harmful. Simple asexual reproduction leads easily into accumulation of harmful mutations in the same way as ailments are accumulating in ageing human bodies. How, then, can the very rare beneficial mutations accumulate in the genome without harmful ones? Species have two ways of doing this. First, if there are enough offspring, some will solely have the beneficial mutations without the harmful ones, but offspring numbers must be very large. Second, with sexual reproduction, even if the genes are polluted with harmful mutations, the lineages can be cleaned of them without losing the beneficial ones thanks to recombination with genes from other individuals. With less offspring, asexual reproduction can be considered short-sighted and unsustainable with deteriorating genes, but perhaps useful to rapidly establish and occupy a new area.

As pioneer trees have fast life cycles and are short-sighted (Weird thinking 3.2), they could initially be thought to suffer less from deteriorating genes and to therefore exhibit more asexual reproduction. This thinking is supported by differences in boreal tree species, of which angiosperms can sprout from trunk bases, produce root suckers, and dominate early stages of succession, while gymnosperms cannot or only rarely reproduce asexually and dominate later successional stages. However, the short-sightedness within the life cycle of an individual tree and in the evolutionary timescale are unrelated, and deteriorating genes are also destructive for pioneer trees.

What, then, could explain the more common asexual reproduction of boreal pioneers? Sprouts are highly effective and gain height rapidly when each day counts during the first few years after a disturbance of moderate severity. However, the most severe fires also kill roots, forcing boreal angiosperms to colonize the area using seed germination. Furthermore, asexual reproduction can be necessary in the shady conditions of a late successional forest, in which tiny seedlings without energetic support from canopy trees cannot grow. The prevalence of sexual and asexual reproduction in species along the fast–slow continuum appears to be complicated.[22] I would expect clearly patterns in a global analysis with asexual reproduction being more common with common but mild disturbances. The boreal pattern with common asexual reproduction early in the succession could be caused by differences between angiosperms and gymnosperms (Chapter 4), rather than directly through the species' life histories.

Weird thinking 3.2 Could interest rates be used to describe the investment strategies of trees?

Accounting for interest rates is essential in long-term economic computations in environmental economics (Figure 9.1) and in silviculture (Chapter 10). With the usual positive interest rates, a given monetary income or expense has a lower present value the further it is in the future. Under a 5% interest rate, for example, an investment in planting seedlings with 100 monetary units should yield at least 105 monetary units more relative to not planting after one year, or 1,147 monetary units more after 50 years to be worthwhile.

Analogously, but with an energetic instead of monetary currency, interest rates can be used when understanding strategies. Trees have positive interest rates similarly to investors. Think of a given tree in a dense self-thinning, even-aged monocultural plantation that receives an energetic boost thanks to which it can grow one extra metre. It makes a big difference whether the tree receives the boost now or in a year's time. If now, the tree is immediately saved from the fierce competition and, with positive feedback, is able to photosynthesize much more and gain a dominant position in the canopy for decades. If in a year's time, it could die during the year while waiting for the boost. Therefore, a boost now is better than later. If extra energy of 100 units now is comparable to 105 units in a year's time in the remaining lifetime reproduction or fitness, then the interest rate of the tree in that situation would be 5%. After maturity the expected lifetime reproduction does not depend only on mortality

continued

---

continued

---

during the period of examination and potential changes in competitive position, but also on reproduction during the period. Similarly, as when considering the term "cost" (Weird thinking 6.2), in addition to energy, also nutrients or even water can be considered. How much more valuable is it for a tree to get a nitrogen dose now than in year's time? Individual trees have variable interest rates during their lives. The urgency is obvious in a crowded self-thinning stand where trees fight for survival, take high risks, and have a high interest rate. If one of these trees is lucky and all its neighbours die, the urgency is over, the interest rate drops, and risk-taking is no longer wise. Tree species have variable interest rates, and the fast–slow continuum[12] can be understood based on fast species having more urgency and higher interest rates. Fast species with higher interest rates have lower wood density to reduce the construction cost at the expense of future maintenance costs,[23] and have higher windthrow risk[24] and fewer defences against pathogens.[25] Interest rates could be used to develop a mathematical theory on the cost incurred by trees for waiting.

A potentially fruitful perspective for understanding natural disturbances is to focus on disturbance probabilities relative to the time since disturbance. To illustrate the potential importance of this perspective, I focus on fire. A forest fire is very unlikely to ignite soon after a previous fire has fully extinguished, as the fuels that burn easily have already burned out. Development during the upcoming years can take many paths. Sometimes the plant species living in the burned area are very flammable, such as dried herbs during the dry season, and their flammability is further boosted by canopy openness, which accelerates fuel drying, increases wind speeds adding to fire intensity, and often also implies an abundant and continuous cover of these fuels. If this period with high flammability is followed without subsequent fires by canopy closure and shady conditions, which cause both rapid litter decomposition and high fuel moisture, then the fire probability will decrease after a couple of years has passed since the previous fire. This can be a typical pattern in the semi-arid tropics, for example, which, after a destructive fire, are occupied by grasses (i.e. species of the Poaceae family). With these grasses, either only the above-ground parts die or the entire plant dies and then dries. They can potentially reach a very flammable state depending on weather conditions, and fires can then kill all the tree seedlings. However, if trees are able to occupy the site they may outshade these light-demanding plants, in which case both the probability of fire and also the intensity of a potential fire decrease dramatically. Many ecosystems have an opposite dynamic. If the main fuels are litter, such as twigs, dead leaves, and cones, which accumulate on the forest floor due to climate factors and litter quality that do not facilitate litter decomposition, then the probability of fire can escalate considerably with increasing time, even after centuries. This is the pattern often described for the arid southwestern United States.

The fire probability trend since the last fire can be useful for understanding not only natural disturbance dynamics but also for comprehending why forests have the structure that they do in biomes around the world, how likely they can tip into another ecosystem, and why the borders of various ecosystems can be either gradual or abrupt. When the fire likelihood increases with increasing time since the last fire, the fire intervals vary relatively little. In the opposite case, when the probability of fire decreases, sudden spatial changes, such as abrupt borders between a forest and an open ecosystem, and temporal changes, like tipping from an alternative stable state to another, become more common. In these cases, an open ecosystem can be transformed into a forest if the very frequent fires due to random variation do not occur for a few years. This could then result in a closing canopy and development into a humid forest in which fires are rare, and if they do occur, have very low intensity and severity. The opposite development is widely discussed nowadays, as it is usually considered negative and could be triggered by, for example, increasing temperatures due to climate change. A forest that has existed for millennia could encounter an exceptional fire followed by a high probability of fire leading to subsequent fires and a more or less permanent conversion into an open ecosystem.[14] A similar pattern triggered intentionally by people has been a major factor in transforming large forested areas into pasture and croplands (Chapter 9).

The examples in the two previous paragraphs concerned fires, but similar trends can be discussed and described for any disturbances, and their interactions are possible. For example, herbivory by large mammals in African open ecosystems can, similarly to fire, be more likely in more open systems. Therefore, an event such as disease and the high mortality of large herbivores can tip the ecosystem into a more closed state that does not revert back to an open ecosystem even after the epidemic is over.[15] However, herbivory is not always considered a disturbance due to its more continuous nature compared to something like a fire.

When examining global disturbance patterns, curious differences appear within a biome. For example, North American boreal forests have more large-scale natural disturbances compared to Eurasia despite the similar climates, as I described above. A possible mechanism explaining this difference operates on an evolutionary timescale. If initially there is a difference in the disturbance regimes between continents, then fast tree species thrive in the regime that has more frequent disturbances, which favours the onset of reproduction at an early age, little investment in surviving other disturbances and stress factors, and inevitable death at a relatively young age. This could explain why North American boreal forests exhibit more stand-replacing fires[8] and insect outbreaks. Theoretically, a North American tree species would do badly in the current Eurasian disturbance regimes, as these species begin deteriorating when succession is still in its early phase, and correspondingly Eurasian trees in North American forests would be killed before they have started reproducing significantly. The reason behind the initial trigger for the divergence of the Eurasian and North American disturbance regimes remains unclear. Similar mechanisms in varying topographic conditions could explain why tree heights and biomasses are surprisingly small in steep landscapes (Chapter 8), as frequent landslides favour the fast strategy, making other disturbances more frequent as well. These same mechanisms explain longevity patterns in animals and the cycle cost analysis in engineering (Weird thinking 3.3).

Weird thinking 3.3 What do fast trees, lions, and Karelian smoke saunas have in common?

Soon after the collapse of the Iron Curtain and even sooner after finishing my secondary school in 1994, I participated in a hiking guide course organized on both the Finnish and Russian sides of the previously largely closed border. When chatting with a group speaking a mix of closely related Karelian and Finnish languages about a newly built smoke sauna in a village on the Russian side, I was told that it was constructed using fast-decaying *Populus tremula*. Decay-resistant *Pinus sylvestris* was abundant in the surroundings of the village, but I was told that rotting would not be an issue as the sauna was likely to burn down before decaying significantly, and that *Populus* logs are lighter to carry and lift. Smoke saunas are energy-efficient to warm, as they lack chimneys and the smoke rises through the room and exits from a small vent. Therefore, the risk of an ember landing on the wooden surfaces and igniting the building is significant.

Larger mammals live longer,[26] but lions typically live just over 10 years,[27] much shorter than would be expected for their size. Lions live a dangerous life with a high mortality risk caused by other lions and aggressive prey such as African buffalos. However, even in zoos lions age much faster than humans. Early ageing is not an issue for most individuals in the wild because they are likely to be killed by another lion or by an aggressive prey.

Pioneer trees that dominate forest ecosystems in the early phases of succession typically have a fast strategy. In natural succession, they are normally overtaken by initially slower-growing trees but have some years to reproduce effectively and disperse their seeds to other recent gaps. Exceptionally, in a garden with no overtaking neighbours, they die anyway due to decaying wood, similar to the lions in zoos. In natural forests, these pioneers rarely survive to the age in which decay begins causing problems.

Smoke saunas, lions, and pioneer tree species are likely to receive their deadly blows from a burning ember, aggressive buffalo, or taller neighbouring trees relatively early, so pioneer trees do not invest much in decay-resistant wood or disease resistance. Thanks to their short-sighted strategies, the smoke sauna is lighter to build, the lion is a faster runner, and the pioneer tree is quicker to grow. Carpe diem.

Natural disturbances have been fundamentally important in plant evolution. It is easier to contemplate how trees would look like without fires than without wind, as fires are not present in all ecosystems while wind is present everywhere. Understanding the role of natural disturbances is challenging also because disturbances, by definition, cause first rapid and then slower changes in growth conditions. In some ecosystems, a short disturbance interval may influence not only evolution directly relative to the disturbance but also the habitats in which a species has evolved. Some species may find a refugium from fires on rocky outcrops that are so dry and nutrient poor that a significant fuel build-up and spreading of fire are not possible. Fire refugia can also be found in the other direction, such as from wetlands where fuels are so humid that they do not burn easily. Such evolutionary pressures could direct tree evolution to withstand very variable soil moisture levels. In between fires, a species could be more abundant in normal dry soils, but just after a major fire it could spread back from its wetland refugia. Then, the species could grow well or even better in wet soils despite most of its biomass being in dry soils when averaged over the disturbance cycles and despite botanists describing the species as normally found in dry soils. In this case, a small but important bottleneck population needs to compete with wet soil specialists. It is possible that *Pinus sylvestris* in boreal Europe has had such a dynamic in its evolutionary history. It is predominately a species of dry soils, but, if looking back at their ancestors were possible, a significant proportion of the individuals may have grown in wetlands. Textbooks describe the physiological optimum for species—the conditions in which growth is peaking—and the ecological optimum in which the species thrives the best when competing with other species. These concepts are based on stable situations. Perhaps a "disturbance ecological" optimum could additionally be discussed to describe the range of conditions in which a species does best in the presence of natural disturbances.

4

The Mystery of Persisting Gymnosperms

If gymnosperms were rarer, they might be called living fossils. Why they are not rarer, or even extinct, is mysterious and fascinating, but seldom discussed.

Most trees, or even all of them based on many definitions of trees (Chapter 2), are either angiosperms or gymnosperms. These are precise taxonomic terms reflecting the relatedness and common evolutionary history of species within these clades. In this book, angiosperms only include eudicot trees (Chapter 1), which are often called broad-leaved trees, but this can be misleading as some of them, such as in the genus *Casuarina*, have scales rather than broad leaves. Another term that I avoid is "flowering", even though all angiosperms, including nontrees, are often called flowering plants. Using the term "flowering trees", however, could make people envision only species with large flowers. Most of the 300,000 angiosperm species,[1] including nontrees, are herbaceous and approximately one-quarter are trees.[2]

Gymnosperms instead are said to not have true flowers. The term "gymnosperm" can be used almost interchangeably with "conifer" when the focus is on trees, as nearly all gymnosperms (trees as defined in Chapter 1) are also conifers, with the exception of *Gnetum gnemon* and *Ginkgo biloba* from the once species-rich order Ginkgoales. When considering nontrees as well, the total species number is over 1,000, and most of these are trees,[3] but a fair number of shrub species also exist. Their species richness is the highest in southern China and southeast Asia.[4] Climatically similar regions in Africa and the neotropics do not have almost any gymnosperms, but southwestern North America is another hotspot thanks to the genus *Pinus*.

The picture is very different when not focusing on species richness but on the relative importance of gymnosperms. In current natural forests, gymnosperms are rare in the tropics, including southeast Asia, and absent in large areas of tropical rainforests. Some drier fire-influenced tropical savannahs on poor soils are exceptional for the regions because they mainly have gymnosperms of the genus *Pinus*, such as in Thailand and

Trees and Forests of the World. Markku Larjavaara, Oxford University Press. © Markku Larjavaara (2026).
DOI: 10.1093/9780197757109.003.0004

Nicaragua. The western and partly southeastern contiguous United States are gymnosperm-dominated, while the eastern and northeastern regions have angiosperms. These patterns are less clear in the temperate biomes of Europe and China. The circumboreal forests are mainly dominated by gymnosperms. The number of angiosperm species is nearly 100-fold relative to gymnosperms. Gymnosperms are distributed on all continents and constitute over 39% of the world's forest area.[5] The total global biomasses of angiosperms and gymnosperms have not been estimated directly, but based on the biomasses of all trees in all biomes,[6] I estimate total angiosperm biomass to be five-fold relative to gymnosperm biomass. However, gymnosperm wood harvesting for industrial purposes significantly exceeds that of angiosperm wood[7] due to its importance in plantations.

Advancements in biological evolution are analogous to those in technology or science. Most of the time advancement is slow and occurs during brief periods of rapid change,[8] which may be triggered by expeditious shifts in the surrounding conditions or in a smaller population of organisms, developers, or scientists. In biological evolution, these periods of rapid change occur within small peripheral populations or following rapid environmental change,[8] which may then result in novel structures or processes. Sometimes these changes should not be called advancements but simply modifications that may be advantageous in the environmental conditions under question, but not in the context of the whole evolutionary history, and may include changes towards simpler structures.[9] On average, changes are towards the better, and the novel structures or processes could be called "innovations". Clades with more "innovations" can be considered more "advanced".

The ancestors of modern angiosperms evolved from gymnosperms and have several evolutionary innovations. The most obvious are the differences in flowers and seed development. Angiosperms produce true flowers and have fast fertilization and enclosed seeds. In contrast to gymnosperms, angiosperm flowers are often pollinated by animals and their seeds are dispersed by animals. This could, in part, explain the global pattern of angiosperms being more common in warmer climates that are conducive to flying insects.

For the topics discussed in this book, more important than the differences in reproduction are the differences in wood and leaves, which are more complex and enable faster transpiration and carbon assimilation in angiosperms. Most gymnosperm wood consists of long tracheids that both transport water and dissolved substances upwards and provide structural support, making it highly suitable for producing strong paper.[10] Instead, nearly all angiosperms have specialized vessels for water transport in addition to fibres providing mechanical support. These vessels can be either spread relatively evenly, in which case the wood is called "diffuse porous," or they may form a ring of wider vessels developed early in the spring with much less tissue conduction forming later in the growing season, in which case they are called "ring porous." However, these are once again not distinct classes, but intermediate anatomies are also common. Because water transport efficiency scales to the fourth power of the conducting tube diameter,[11] angiosperms with larger vessels potentially have a massive transporting capacity for a given stem cross-sectional area relative to gymnosperms. This quality is particularly valuable for woody climbers or lianas that are mechanically carried by self-supporting trees. Unsurprisingly, all lianas are angiosperms. The exception are lianas in the genus *Gnetum*, which are the only gymnosperms with vessels in their wood.

Vessels are often considered the most fundamental evolutionary innovation when angiosperms first evolved. However, specialization increased not only wood cell numbers but the increasing complexity in leaf veining structure in early angiosperms also made a large difference.[12] Gymnosperms typically have only one or several similar parallel veins, but angiosperms have veins of various sizes, potentially better providing the needed mechanical support and efficient water transport.[13]

If someone who has not studied botany is asked to describe differences between angiosperms and gymnosperms, or their equivalents in colloquial language, the answer is not likely to include vessels in wood or veins in leaves. Instead, the focus is likely to be on the small needles or scaly leaves of gymnosperms relative to the broad leaves of angiosperms. People from boreal or northern temperate biomes with only few evergreen angiosperms may add that gymnosperms are evergreen and angiosperms deciduous. Next, they would describe how gymnosperms have a narrower crown with shorter branches.[14] People from the Mediterranean region, with plenty of evergreen angiosperms, or from the Siberian deciduous *Larix*-dominated area, would just talk about the small leaves and conical crowns of gymnosperms but not of evergreenness. Gymnosperms are naturally so rare in the tropics that if asked about gymnosperms, laymen from the region would probably start thinking about exotic species plantations or pictures from colder regions.

Foresters worldwide value gymnosperms, and if asked would add a description of their straight trunks that rarely fork and therefore produce high-quality timber even with sloppy silviculture. Instead, angiosperms are more responsive to changes in light conditions. The value of their trunks can decrease due to large branches, secondary trunks, and epicormic shoots along the previously branchless quality trunk when individual trees are exposed to additional light after a thinning. Epicormic buds are typically common at the bases of angiosperm trunks cause sprouting from stumps, which is problematic to foresters wishing to grow gymnosperms, but concurrently allows coppicing if rapid angiosperm wood or leaf biomass production is the objective.

Various perspectives have been considered when comparing angiosperms and gymnosperms in scientific literature. Gymnosperms average 14%[15] to 20%[16] lower wood density than angiosperms do, explaining why their timber is called softwood. Low wood density is normally associated with a rapid life cycle, fast growth, and high mortality[17] due to high risk-taking.[18] However, gymnosperms, contrary to what is expected for low-density trees, grow slowly especially when young,[19] have wood that is highly resistant to decay,[20] and reach older ages than angiosperms.[21] Gymnosperms have a slower life cycle but still mysteriously have low-density wood.

How can the varying descriptions by laymen, foresters, and scientists regarding the differences between angiosperms and gymnosperms be understood? I argue that those explanations that are not based on the fundamental evolutionary innovations of early angiosperms on the complex and variable wood and leaf structures with vessels, and on the variable-sized veins, may be correct but are still not sufficiently satisfactory. For example, stating that gymnosperms do better in cold climates because large-diameter angiosperm vessels are vulnerable to freezing and thawing is misleading, as structure size can change rapidly in evolution contrary to increasing complexity. If advantageous, angiosperms could have evolved wood with small vessels for cold conditions, like gymnosperm tracheids. I try to base my explanation on the more complex wood and leaves of angiosperms, but unfortunately many of my arguments are only tentative and a couple of key steps in the reasoning are missing. However, I hope that my speculations encourage others to develop more sound theories and collect data to test them.

Because gymnosperms lack vessels, they need a larger cross-section in their branches and young stems for a given leaf area. Therefore, they have lower wood density so that they can build a given cross-section area with lower biomass investment. Larger stems or trunks are not influenced directly by the weak water conductance of the vesselless wood, as they need to have thick trunks for mechanical stability (Chapter 7), but their wood density may be low because altering wood density significantly during development can be difficult. Despite the low wood density, the diameter needed for sufficient water transport is so large that the material investment is greater and young gymnosperms therefore grow more slowly. The slow initial growth may influence the entire life cycle, as trees that have survived the challenging early stages optimally lengthen their mature phase and reproduce for longer and therefore invest more in traits that reduce mortality, such as decay-resistant wood. Similarly, they must invest more in branch wood for a given leaf area and therefore may be able to carry the winter snow load with a smaller additional investment in wood; consequently, they are more likely to have evergreen leaves.

Understanding how the other angiosperm innovations, such complex vein structure and other potential differences in leaf development and structure, influence current overall gymnosperm structure and distribution is challenging. However, it seems likely that something in the construction of gymnosperm leaves potentially links to the simpler vein structure that makes building large leaves expensive and renders the construction costs relative to maintenance costs high. When maintenance is cheap relative to construction, it is better to prolong leaf lifespan by adding compounds reducing herbivory and increasing the benefits of evergreenness. Slow leaf turnover slows whole-tree turnover similarly, as the slow initial growth rate is due to vesselless wood. It is unclear why gymnosperm leaves are small, but their small leaves boost ecosystem photosynthesis, all other aspects being equal, as the lower leaves are exposed more to intermediate levels of light than under a canopy of large leaves. For these same reasons, gymnosperms with small leaves are less efficient at blocking light from neighbouring trees of the same or of other species and therefore focus on their own height growth instead of spreading their crowns to obstruct their neighbours. This, together with the need to carry snow and ice loads as an evergreen tree, has led to the narrow conical crowns that are typical in most genera in the important gymnosperm families Pinaceae and Cupressaceae. The water transport issues could also make long branches inefficient, thus narrowing gymnosperm crowns. In addition, other factors unrelated to angiosperm–gymnosperm differences could influence crown widths (Weird thinking 5.2).

Imagine a forester having worked on plantations with angiosperms all their life. They would be excited when discovering gymnosperms, as most deviations from angiosperms are for the better. The slow growth when young is a drawback, but the extra wood that is needed by the tree to transport water can be used as timber. The benefits of evergreenness are more significant (Sceptic's question 2.1), ranging from the decay-resistant wood to the narrow crowns of trees that have taken a game theoretical approach of minding their own business and not focusing on blocking others, which lowers overall growth and reduces timber quality.

Gymnosperms are better in many ways for foresters producing timber, but how have they been able to compete with angiosperms and persist in nature for over 100 million years before they abruptly began being planted around the world's forests in a kind of mutualism similar to that with domestic animals? They were not just persisting, but doing well if the measure is proportional to the number of species, better when the focus is on the proportion of global biomass, and even better when considering the forest area that they dominate. The evolutionary innovations of early angiosperms increased the complexity of wood and leaf tissue, enabled more efficient functioning, and did not preclude simpler, less efficient structures in evolutionary timescales. Not only do gymnosperms persist, but so also does the mystery behind why they endure.

PART II
LIMITS TO TREE SIZE

5

Tree Structure

Imagine being an engineer and having to design a supporting structure for solar panels in a dense forest. Some engineers might think that reaching the top of the canopy is too difficult, and would simply develop a low frame and invest the funds saved by not building a high structure into installing more panels. Certain panel types could yield significant energy levels even in shady understorey conditions, but such panels are at risk of being damaged by falling trees or branches. The most inventive engineers might consider attaching the panels to the tallest trees and installing electric cables running down the trunks to the ground. This could be a cost-efficient plan, but with numerous challenges. The top branches are so thin that attaching heavy panels would be very difficult, and trees or lianas growing in the tropics could overtop them. In addition, the supporting tree might fall and destroy the panels. Most engineers would go for the most obvious solution: a tower supporting the panels above the canopy. However, even this tower could be constructed in numerous ways. For example, the mast could be a metal lattice or a tubular structure or perhaps a solid wooden or concrete pole. All of these structures could be self-supporting or secured with guy wires, buttresses, or extensions increasing the base diameter.

These challenges faced by engineers are similar to what evolution has tackled. There are many differences, but possibly fewer than initially pictured. The panels installed on the ground resemble shrubs, and those attached to the treetops resemble lianas. Shrubs may be damaged by falling debris.[1] To have a positive energy balance, they need to be shade-tolerant in forest ecosystems, or they may be energetically starving due to a sudden change from their less shady past. Liana leaves and twigs are lighter than solar panels and better able to climb the finest branches on the tops of tree crowns. However, they are structural parasites and must tackle challenges that do not affect self-supporting plants at evolutionary timescales.[2] Firstly, they must be able to grow upwards in the shade of a tree. Secondly, they are not only vulnerable to their own pests and pathogens or to other deadly disturbances but also to those of the host tree. If a liana shades and kills its

Trees and Forests of the World. Markku Larjavaara, Oxford University Press. © Markku Larjavaara (2026).
DOI: 10.1093/9780197757109.003.0005

host tree, it will initially have increased the light available for itself, but the situation will inevitably lead to a catastrophic collapse.[3]

Initially, the engineer seemingly has more options for designing a self-supporting tower than what evolution has created. Evolution has apparently been unable to innovate more complex structures, which is possible for engineers. However, when considering the above-mentioned options for a self-supporting tower, similar structures can be found in tropical rainforests and in some other biomes. Most bamboos are hollow, resembling tubular poles. Lattices or web-like structures are rare, but stranglers in the genus *Ficus*, which germinate in the crowns of other trees, grow aerial roots down to the ground and slowly strangle their hosts to death while simultaneously increasing their own ability to remain erect unassisted. Subsequently, this can closely resemble lattice towers. The aerial roots of these same trees can also look much like guy wires even though they are vertical. The stilt roots common among mangrove species resemble guy wires even more, as they are at a nonvertical angle. Finally, many tree species develop impressive buttresses, especially in tropical lowland rainforests (Sceptic's question 5.1).

Sceptic's question 5.1   How are buttresses beneficial?

Based on engineering theories, buttresses strengthen the trunk base and the connection of the trunk to the root system.[16] This has long been established in ecological literature as the main reason for the existence of buttresses, and datasets have shown that the windward sides of trees develop larger buttresses,[25] which is understandable because a bending trunk much more effectively pulls up roots running on the surface than pushes them downwards. This has led ecologists to hypothesize that buttressed trees are more likely to snap above the buttress rather than to uproot.[26] The lack of support for this hypothesis[26] has suggested that buttresses do not function biomechanically and encouraged scientists towards creativity and to ponder whether, for example, buttresses could be useful in trapping wind-blown leaves to make more nutrients available.

However, we should not compare a buttressed tree to an identical tree lacking a buttress. Why would a buttressed tree waste resources and develop an otherwise identical trunk base and coarse roots if the buttress already adds strength? Instead, we should compare a buttressed trunk to an unbuttressed trunk with a given construction or maintenance cost or to other measures that are meaningful in tree evolution, because a buttressed trunk base adds surface area and maintenance costs relative to construction costs, for a given wood density. Focusing on the anchorage is complicated by soil characteristics, but the star-like cross-section provides more strength along the trunk for a given construction cost but also incurring a higher maintenance cost for a given strength if much of it is caused by the trunk surface. In this way, comparing two trees that are otherwise identical, but one is buttressed and the other is not, is like comparing trees with identical trunk diameters but variable wood densities, which may lead to untrue conclusions.[22] We should be careful when making comparisons to answer other ecological questions, as we must be sure that the comparison is meaningful. It makes more sense to compare trees with similar construction or maintenance costs, or comparable trunk strength.

Evolution is apparently able to develop structures that engineers also use to design self-supporting towers, but these structures are only marginally important globally because they are inferior to the normal solid pole structure of most gymnosperms and angiosperms. Engineers would not raise solar panels above the canopy with a solid metal pole. Therefore, there are clear differences between an optimal self-supporting metal structure and a tree. Could these differences be caused by the transportation and storage needs of trees? This is unlikely. Biomechanics and the need to resist bending and breaking forces likely determine tree trunk dimensions and heights in most cases (Chapter 7). Most large trees have metabolically dead heartwood in their trunk cores, which only serves to increase strength and does not contribute to transportation and storage. When once again comparing to human-designed structures, trees appear to arrange their transportation and storage needs similarly to a solar panel-powered streetlamp with a tubular metal pole, with the cables and batteries taking up only a small portion of all the space available within the pole. The pole is therefore mainly designed for rigidity, and the need for cables and batteries does not influence its overall form.

Before discussing the potential reasons why engineers typically design tubular or lattice structures while most forests are dominated by trees with solid trunks, let us first examine the trunks like an engineer would. If a load is attached to a hanging wooden pole, the pole can be assumed to stretch or lengthen slightly and roughly proportionally to the weight of the load. Also, as expected, doubling the cross-sectional area doubles the size of the load that causes a given lengthening independent of the cross-sectional shape. However, the situation is more complex with upward wooden poles or growing trees. When wind causes drag in the crown, the trunk bends and gravity causes additional bending moments.[4]

To understand how the trunk resists these moments, it is useful to assume a neutral axis running lengthwise in the middle of the trunk, in which neither compression nor tension occurs. Wood on the convex side of the trunk is under tension, and the fibres resisting the bending moments stretch while correspondingly the fibres on the concave side shorten. The further away the fibres are from the neutral axis, the more they resist a given bending of a trunk, for two reasons. First, the further away they are, the greater the countermoment that a given fibre causes with a given change in length, as the distance to the neutral axis is equivalent to the lever length in normal moment calculations. Second, the further away a given fibre is from the neutral axis, the more its length is impacted by a given level of bending. Therefore, doubling the distance of the fibres to the neutral axis leads to a four-fold resistance at a given bending. When the focus is not on the countermoments with which the trunk resists a given bending, but on maximal bending just before breaking, doubling the distance of the fibres to the neutral axis only doubles the bending moments that the trunk can resist, as the maximal bending decreases with increasing diameter.

Increasing the distance of the fibres to the neutral axis strengthens the structure without additional materials, explaining why engineers favour tubular and lattice structures. Palms have denser wood in the outer layers of their trunks, and bamboos normally have hollow cores. They therefore attain greater strength for a given material, which calls to question why these groups of plants are not more common. Two issues appear to be the main reasons for the dominating conical solid wooden trunks of angiosperms and gymnosperms. First, trees must grow, which means that a given structure at one point in time is not only adapted to that point in time; that is, part of that tissue was already forming a much smaller individual in the past and most of the tissue will be part of a much larger individual in the future. During their course of evolution, trees have not been optimizing their structures at a given point in time but are rather balancing the value of the structural tissues they produce for all future moments of time. Some species and individuals are short-sighted in certain circumstances, placing more weight on the near future, while others are far-sighted (Weird thinking 3.2). Unlike most engineered structures, which have optimized the production costs and benefits from the end product, tree development more closely resembles structures that are used during the construction phase and are never completed.

The second reason why most trees, both angiosperms and gymnosperms, have similar solid tapering trunks has rarely been discussed and is more challenging to comprehend. Woody plants seem to be built not only to minimize the materials needed to construct the structure over their entire life cycles but also to minimize the surface area of these structures. The picture is not very clear but can be understood by examining it from numerous angles and seeing a hazy yet similar image from each. First, empirical studies attempting to quantify energy consumption (i.e. respiration), which is unrelated to growth, have shown that this maintenance respiration scales well with the surface area of the trunk or branch section.[5] Plants therefore benefit from a lower surface area, as this reduces their energy consumption. Second, the trunk and branch bark must be built, which causes additional construction costs. Third, we may focus on function and examine the reasons why surface area causes extra costs. From a tree's perspective, many biotic and abiotic dangers strike from the surface, and from this viewpoint it is logical to avoid additional surface area. Trees can use insulating bark to protect themselves from hot fires[6] and toxic bark to guard against animal herbivores.[7] Most pathogens also attack from the surface, and active defence or preparation for defence against these pathogens increases a tree's energy consumption. Avoiding structures with low biomass and high surface area also explains the rarity of multistemmed trees, despite this adaptation initially seeming like an excellent insurance against complete destruction (Sceptic's question 7.2). Similar patterns, such as the self-thinning law with smaller biomass per unit area with smaller trees (Chapter 6), may be caused by small individuals that have higher surface-to-volume ratios and therefore higher maintenance costs per unit volume or mass. Strangely, it seems that a large surface area is a smaller problem in warm conditions (Weird thinking 5.1). From an engineering perspective, the problematic nature of a large surface area could be compared to an expensive coating that has to be used to cover a structure. Above a certain coating price, a solid circular structure would become the optimal tower structure.

Weird thinking 5.1 Why do buttresses only occur in the tropics?

It is peculiar that unusually formed woody plants without a simple conical stem, and therefore a larger surface area relative to volume, occur more typically in tropical lowland rainforests than in other forests of the world. These structures include stilt roots and buttresses. Some tropical lowland rainforest species even have a star-like cross-section as their whole trunks and not just as their base like the more common buttressed species. All these deviant structures disappear rather abruptly when either climbing up a mountain slope to a cloud forest or when travelling away from equatorial latitudes.

Fires may partly explain the disappearance of these supporting structures away from the lowland tropics, as insulating a large buttress with thick bark may be overly expensive. It may be cheaper to keep the circular cross-section at the base and to just thicken it and any adjacent roots to increase mechanical strength. However, fires increase only away from the equator, but not up mountain slopes. Fires cannot explain the absence of buttressed trees in tropical mountains or in ecosystems without fires far from the tropics, such as humid temperate forests, or in wetlands with no distinct dry season. Another explanation is based on biomechanics. Could acquiring nutrients and water be so effortless in tropical rainforests that only a small root system is required, which then does not provide enough biomechanical support, leading to the need for buttresses or stilt roots? This is unlikely, as coarse roots close to the trunk base are clearly massive for biomechanical support, and trees do not seem to get this support as a by-product anywhere on Earth. The most likely explanation is based on surface-to-volume ratios. Not only are these trunk base structures providing mechanical support more common in lowland tropics, but many structural plant groups with exceptionally large surface-to-volume ratios, such as lianas and strangler figs, are also typical only in tropical lowland rainforests. Construction costs, maintenance respiration, the energetics of defence against tree pathogens, and the corresponding processes of these pathogens are all temperature dependent. The higher surface-to-volume ratio is likely not as problematic in forests with the highest temperatures and rainfall, but I do not have a clue as to why this is.

I think that it is interesting to compare the four taxonomically and structurally distinct groups of large woody plants. Angiosperms (eudicot trees as defined in Chapter 1) dominate most forested biomes of the world, although monocot palms, another angiosperm lineage with a similar age of over 100 million years,[8] are common in the tropics and subtropics yet contribute only a very small portion of total biome biomasses when compared to angiosperms. On the other hand, a much older lineage, gymnosperms, with a history of over 300 million years[9] still dominates the biomass in most boreal and some temperate forested systems, despite the relatively small species number[10] (Chapter 4). The fourth group, woody bamboos, is a much younger lineage with a history of some 50 million years.[11] They are similar to palms, common in the tropics and subtropics, but are only of marginal importance relative to angiosperms. Stereotypically, we may expect newer lineages to have improved design and to gradually replace the more primitive lineages. However, gymnosperms still dominate in the boreal (Chapter 4), and the more recent palms and bamboos are still only of low importance, even in the tropics. Biomechanically, gymnosperms and angiosperms are similar. Their older design relative to palms and bamboos is apparently excellent comparative to the newer structures. Palms appear biomechanically handicapped, and it is surprising that they have ever even evolved. Their trunks provide more strength for a given material investment for a given height, but their inability for secondary growth is a major problem, as without diameter growth they require a biomechanically excessively thick stem when young. Bamboos are even more different relative to normal angiosperms and gymnosperms because they have hollow stems that elongate from each trunk section.[12] They seem to have numerous biomechanical advantages but oddly have been unable to evolve into large, crowned plants. Most, if not all, bamboos appear able to bend to the ground unharmed and therefore have a similar biomechanical strategy as shrubs.[13] Their modest leaf area per stem is probably related to this shrub strategy, as a greater leaf area per stem would cause gravity to buckle the stems.

Trunks are characteristic of trees. Only trees have trunks, but their wood and the branches and roots attached to them are not that different from those attached to the stems of shrubs. Tree trunk structures have converged during evolution, and nearly all angiosperm and gymnosperm trunks are similar biomechanically. Their branching structures vary from angiosperms to gymnosperms (Chapter 4), but also from species to species, and may often be an important clue in identifying trees from a distance. Certain patterns in crown and branching shapes are shared between biomes,[14] habitats, and life strategies, but even ecologically similar species can differ drastically in this respect. This suggests that these alternative crown designs do not significantly influence fitness.

Branch angle is possibly the most noticeable crown feature. In windless conditions without additional weight, such as snow on the branches, the optimal angle would probably be slightly upwards from the horizontal. Such angles lead to nearly as great a spread of the leaves as a horizontal arrangement does, but the leaves are concurrently lifted some distance upwards. A more upward angle would allow for faster replacement of a broken main stem by one of the branches, as they are already oriented upwards. Additionally, the upward angles could lead to greater stemflow, perhaps saving the tree if lightning strikes, as the current may then travel on the bark surface and not in the sapwood. Downward angles are advantageous if temporary snow loads reach significant levels for evergreen species. This is because a branch can resist far greater weights when hanging rather than being in a horizontal position. Similarly, branches that should be as rigid as possible in calm conditions benefit from streamlining in windy conditions, bending parallel with the wind.

The main trunk grows, with old parts hidden beneath newer layers, but a large portion of the branches that grow is shed during a tree's lifetime. Lower branches, initially potentially in full light, may be shadowed by the leaves of the same tree or neighbouring trees, resulting in negative net production, and are therefore better shed (Weird thinking 5.2). However, at least theoretically, it may also be worth keeping energetically useless branches and their leaves alive if they help to shadow plants growing underneath from competing below-ground. On the other hand, a clear, branchless bole may concurrently be beneficial for avoiding leaf browsing, being killed by fire, or even to reduce the likelihood of lianas, flames, or herbivores climbing to the upper parts of the tree. Individual angiosperms seem to exhibit larger variations in branch size and forking structure than gymnosperms do. This angiosperm design may be better when flexibility is needed in crown width. Overall, gymnosperms are apparently not just less opportunistic and flexible in their search for canopy space, but also have narrower crowns in general.[15] This goes hand in hand with their small leaf size, which makes shading their neighbours difficult (Chapter 2). Just as with human games and societies, competition among trees has two extremes: one extreme is to focus on an individual's own growth, and the other is to block the growth of neighbours. I suspect that trees with more vertical canopies probably have smaller leaves and produce heavier and especially taller trunks, as they do not spend resources on trying to block others. Therefore, gymnosperms and angiosperms with narrow crowns, such as in the genus *Eucalyptus*, are popular in plantations (Chapter 10).

Weird thinking 5.2 Why do trees have branches along their trunks and not just on top?

Competition for light is key when trying to understand trees.[34] A pioneer species initially grows fast thanks to its greater photosynthesis in full light and its other pioneer traits (Weird thinking 3.2). A shade-tolerant species is a better competitor in low light. If an engineer is given the task of designing an optimal crown for the pioneer species, they might first come up with the idea of flat crowns with all the leaves arranged in a horizontal layer, with so many leaves that the light levels below it would be too low even for the shade-tolerant species. However, we rarely see such canopies even though they would shade out the competing species. Yes, there are palms with large leaves or fronds only at the top, but they unable to grow leaves elsewhere. Some angiosperms fork several times, reduce their height growth, and eventually develop relatively flat crowns. These can be seen exceptionally in tropical rainforests or savannahs, where fewer leaves along the trunks may help to reduce browsing and damage from wildfires.

The engineer is not considering two issues. First, to have leaves only at the top, growing trees would need to build new branches every year and the older branches would be useless. Second, trees have only a limited ability to place their leaves relative to their other leaves, and a carpet-like single layer design is impossible. Therefore, to create sufficient shade to supress the shade-tolerant species requires some of the pioneer leaves to grow in a light level in which they consume more energy than they produce, which is harmful for the tree individual. Not only do species compete in evolution, but so do individuals as well. A free-rider within the pioneer species would do better and spread its genes by not developing the leaves to supress the shade-tolerant species.

continued

continued

Placing leaves at other heights than at the top has further benefits. Winds are weaker at lower heights, and a given force caused by wind or gravity causes less toppling over moment for the trunk. Leaves cause a full shade only some distance downwards and therefore, with a given leaf area above, it is better to have the shading leaves at a greater distance than to have better light quality (Chapter 2).

Even though clearly it is not realistic for trees to have leaves only at the top, this contemplation helps in understanding the variation in crown length,[14] which has been linked to latitudes[35] and the fast–slow continuum.[36] In the intense competition for light, crowns are expected to be short, as light levels decrease quickly downwards and because upper leaves are more harmful to their neighbours. This is what foresters wish would be the case (Chapter 10).

Roots receive less attention from forest ecologists, as they are only rarely visible. The flow of water with nutrients to the crown begins from the roots. In addition, roots provide anchorage and storage space that is even better protected from most herbivores and other disturbances than the trunk. Picturing the root system as a below-ground mirror image of the above-ground parts, as is sometimes done, is generally a weak argument because many tree species do not develop tap roots[16] that are analogous to trunks. Secondly, root biomass is typically only approximately 50% of the above-ground biomass for shrubs in arid climates and 20% for trees in humid climates.[17] However, some interesting equivalences can be seen. Nutrients are typically more abundant on the surface, but water availability in arid soils increases downwards in a similar manner as light levels increase upwards. If the roots reach a certain depth, abundant water is normally available, similar to full-light conditions for leaves above a certain height. Therefore, the competitive advantage may shift from herbs to trees or shrubs in arid climates without a clear rainy season, but with abundant water in deeper soil layers. Nutrient and water foraging by the roots can be divided into the actual search for nutrients and water and into harming neighbours by nutrient and water uptake that could alternatively be taken up by the neighbours themselves, thereby reducing their vigour and competitive capability.

These questions are parallel to the discussion above concerning crown expansion to block light from neighbours but is possibly even more difficult to quantify, despite small-scale experimental studies having demonstrated root aggression towards their neighbours.[18]

Roots far from the trunk base are subject to completely different biomechanical constraints than branches are, as they do not have to resist forces, such as gravity and wind, but instead must occasionally penetrate very hard soils. However, coarse roots close to the trunk base provide anchorage for the above-ground parts of a tree, and their structure can therefore be partly understood based on biomechanics. As expected from an evolutionary perspective, most wind damage studies report both trunk snapping and uprooting,[19] indicating that they are reinforced in synchrony so that neither is significantly stronger, as unnecessarily excessive strength increases construction and maintenance costs without benefits. Interestingly, trunks are normally nearly symmetrical and therefore resist winds from all directions similarly, but the development of root anchorage is probably very sensitive to prevailing winds. Therefore, I expect a strong wind from an unusual direction to lead to relatively more uprooting.

Trunks, branches, and roots have surprisingly similar wood. Wood has evolved into a material that enables cost-efficiently building structures needed by trees. Wood mainly contains elements that trees obtain from water and atmospheric carbon dioxide, and therefore its synthesis can be thought to be limited energetically or by photosynthesis and not be based on nutrient availability (Weird thinking 2.1). By their common definition, the stems of woody plants contain lignin, a group of complex polymers. As with many biological terms, there is no sharp boundary, but many plant species go through increasing woodiness during their development and may have woody tissue only in their stem bases. Herbaceous plants or herbs, by contrast, typically lack lignin. Both woody and herbaceous tissue have a large proportion of cellulose, which is one of the strongest materials known for its mass when stretched.[20] Lignin and cellulose function in wood similar to how concrete and steel function in reinforced concrete. Lignin provides resistance against compression as concrete does, and cellulose resists tension as steel does. The role of lignin in wood can be understood when thinking of a cotton cloth that is nearly pure cellulose or a herbaceous plant that slumps down when wilting and losing its internal water support under pressure. On the contrary, a piece of wood with lignin retains its shape even when dried. The ratio of cellulose to lignin varies when trunks are not vertical (Sceptic's question 5.2). In addition to cellulose and lignin, wood contains hemicelluloses and a great range of other compounds often called extractives or secondary compounds. Most of them have defensive functions, and their diversity is understandable, as more herbivores are adapted to the common defensive compounds. Developing different compounds has therefore been advantageous. The species specificity of herbivores and pathogens, not only of wood but of all kinds of tissues, is an important component of theories explaining the great number of tree species. Newer and rarer species benefit from fewer herbivores and pathogens utilising them and therefore are less likely to go extinct (Chapter 1).

Sceptic's question 5.2 Why do trees develop reaction wood?

Reaction wood is unusual wood that differs from normal trunk wood in its density, chemical composition, and cell structure. Reaction wood has been studied extensively, as it causes severe problems in the wood processing industry.[27] Reaction wood in angiosperms is called tension wood, and it contains a higher-than-normal proportion of cellulose. Correspondingly, compression wood in gymnosperms has more lignin. Reaction wood is formed in trunks typically during asymmetric growth conditions. The normal thinking is that tension wood is used by trees to pull and compression wood to push back into a normal position. This is a natural answer to this Sceptic's question, as it corresponds to most observations.

However, sometimes reaction wood does not pull or push a form back towards the symmetrical or normal. For example, in conditions with strong prevailing winds from a given direction, gymnosperms develop compression wood on the leeward side of their trunks, which clearly is not an attempt to push the crown towards the wind,[28] as that would be biomechanically harmful. Instead, these trees seem to simply increase lignin-rich compression wood on the side of the trunk that experiences greater compressions. Correspondingly, angiosperm tension wood is found in the rope-like aerial roots of stranglers in the genus *Ficus*.[20] We should not underestimate how trees have been capable of adapting their structure during evolution. Reaction wood in general should be viewed as wood that increases tree structure strength relative to its construction and maintenance costs in unusual, often asymmetric situations. Frequently, it is associated with a corrective change in a tree's posture, but not always. Because cellulose provides more tensile strength than lignin does, tension wood is useful in locations where the tissue is normally under tension. Correspondingly, the lignin-rich compression wood strengthens the overall structure if it develops in locations that are usually under compression.

Most tree species develop heartwood when reaching a large size (Sceptic's question 5.3). Heartwood is metabolically dead tissue, is always found in the trunk core, and is typically darker in colour. Heartwood used to be sapwood when a tree was smaller, and the microscopic structure is therefore similar. However, when the tissue dies, it no longer benefits from active biochemical defence against microbes, leading trees to increase their defensive compound concentrations, which also causes the darker colour. Nevertheless, many species often suffer from heartwood decay even though the heartwood is surrounded by sapwood, indicating that active biochemical defence is superior to passive chemical defence. However, sapwood vulnerability to decay changes when the tree dies, while heartwood vulnerability does not. Sapwood becomes vulnerable to attacks by fungi and other microbes or sometimes by even larger animals. This is why larger logs with a greater proportion of heartwood are more expensive on the timber market.

Sceptic's question 5.3 Why do most old trees have heartwood?

Understanding why trees develop heartwood is important, as it can provide key insights into their overall growth and size. One line of thought suggests that trees must produce new sapwood to replace older sapwood that can no longer transport water effectively. This can be either due to ageing or to the loss of connection with new branches or leaves when older ones are lost through crown rise. The old sapwood is therefore converted into heartwood. This reasoning assumes that heartwood is useless or at least does not consider any potential benefits it might offer. However, this is misleading, as heartwood often constitutes over half of the biomass,[29] significantly contributing to the overall construction costs in a tree's lifetime. If heartwood offered no benefit to living trees, there should have been a strong evolutionary pressure to lower construction

continued

continued

---

costs by reducing heartwood volume. Interestingly, trees and liana stems have similar storage and transport requirements but differing biomechanical needs. Lianas, which are biomechanical parasites, have much thinner stems for a given leaf area than trees[30] because they do not have to support their crowns and do not build a massive heartwood core within their stems. This contrast suggests that heartwood is valuable for trees for biomechanical reasons.

A trunk of similar dimensions and similar biomechanic properties could be composed entirely of sapwood. Two possible explanations exist for why this is not typical. According to one line of reasoning, the central parts of trunks cannot have metabolically active, live tissue because oxygen transport, required by living cells, is challenging to achieve in the inner parts far from the surface. This thinking is based on the premise that if oxygen were available in the core, it would be better for trees not to have heartwood. However, many species develop heartwood even in relatively thin branches[31] with a much shorter distance to the surface. It raises doubts about the entire role of oxygen in heartwood formation. According to the second and most plausible line of reasoning, trees have evolved to develop metabolically dead heartwood to have tissue that adds strength without causing maintenance cost. However, trees must add secondary compounds to their heartwood to deter pathogens. Thus, once again, there is a trade-off between construction and maintenance costs, and I would expect long-lived species on average to have more strongly protected heartwood.

Sceptic's question 5.4 What can we understand from wood structure based on water transport?

To understand general wood and tree structure, a large group, perhaps the majority of scientists in the field, favour a perspective that is based on understanding water transport in the sapwood from the roots to the leaves. For example, according to one theory, the overall wood density variation is driven by the need for trees to resist implosion, or breaking of the conduit walls. Based on this reasoning, high wood density is needed to better resist implosion, which is needed in drier climates because the negative pressures in the conduits increase during droughts. However, only a small portion of all angiosperm wood tissue contributes to water transport, and we can ask why wood would need to be denser in general, and not just the conduit walls.[32] I believe that trees in drier climates have higher wood density because they compete less with their neighbours for light, but instead, those individuals that have passed the risky seedling phase involving mortality from fire and herbivores thereafter have a far-sighted investment strategy (Weird thinking 3.2).

Another example of the challenges in explaining wood structure based on water transport is the theory that the evolution of wood balances the efficiency and safety of water transport. Here, safety refers to the risk of embolism in dry conditions that hampers water transport, while efficiency is concerned with the "rate of water transport through a given area and length of sapwood".[33] The conceptual problem in this expected trade-off is that increasing conduit concentrations in the wood would increase efficiency without influencing safety. Even more space for conduits with a given construction cost could be achieved with lower wood density. Therefore, the weak correlation in a global test of the theory[33] is logical.

Research on water transport is challenging to link to angiosperm wood structure in general. Instead, scientists should try to understand the costs of constructing and maintaining only those structures that contribute to water transport. Unfortunately, that would also be tricky. Understanding whole trunks, comprised not only of sapwood but also of heartwood, is even more hopeless based on water transport, as heartwood does not transport water.

Forest ecologists or tree physiologists, with a focus on wood, commonly have either a water or biomechanical perspective. I argue that the water perspective is useful for understanding water-conducting structures, but the mechanical perspective is more suitable for comprehending wood in general (Sceptic's question 5.4). The mechanical properties of wood can be described with two variables. Engineers, who normally do not want the structures that they design to change shape by more than a few millimetres, are mainly concerned with the modulus of elasticity. It describes how stress (i.e. force per unit area) causes strain when applied to a material. However, for trees, and their fitness in their evolutionary history, it is unclear whether greater bending has been good or bad. Increasing bending increases streamlining and reduces wind drag, which may allow a tree to obtain support from other trees and may help it resist short gusts during which the crown sways in the wind without breaking. On the other hand, the more bending occurs, the more likely the crown will be damaged through contact with neighbouring trees, and gravity causes greater loads. The load results not only from fresh mass of the tree but also from epiphytes, water, ice, and snow. The modulus of elasticity is influenced significantly by the microfibril angles in the cell wall, causing either rigid or flexible wood. However, both of these have a similar modulus of rupture, which is more important for trees. The modulus of rupture is the maximum stress that a material can resist and is the strength of the material in bending. It is difficult to envision a situation in which a higher modulus of rupture would not be better for the tree.

Wood density, typically defined in ecology as dry mass per fresh volume, is a main wood trait influencing the modulus of rupture to which it scales roughly proportionally.[21] Because denser wood requires more material per unit volume, dense wood has traditionally been thought to be beneficial through greater strength and lighter wood due to lower construction costs. However, this thinking is based on a comparison of trees with the same diameter but variable wood densities. This is misleading, as often diameter is not a meaningful measure for trees.[22] Instead, if we focus on a comparison of trees with variable wood densities but with equal wood construction costs (i.e. biomass), we notice that with lighter wood for a given construction cost strength can be greater. This can be understood based on the distance of fibres to the neutral axis. Low-density trunks have fibres further away from the neutral axis, which therefore contribute more to trunk strength.

This leads to the question of how higher wood density is advantageous. Lower wood density for a given wood construction cost involves more bark construction, with potentially differing construction cost per mass relative wood and higher maintenance costs linked to the larger surface area. Therefore, constructing a trunk with a given strength involves more material investment with dense wood, but will be cheaper to maintain. Short-sighted pioneer trees often have low wood density,[23] which is understandable due to the cheapness of construction relative to maintenance costs. They can achieve a given strength with cheaper construction costs, and the cost of maintenance in the upcoming decades is less significant for a short-lived tree species.

Wood density variation within a given trunk can be understood based on the same principles and the variation between species. When a tree has an urgency, such as due to a difficult position that will likely cause it to be outcompeted by its neighbours, it produces lower-density wood that helps it to grow in height without risking an excessive risk of snapping its trunk. However, the penalty of this is increased maintenance costs for the rest of its life due to the larger circumference. In the opposite case, when a tree has secured its position in the canopy, most species will increase their wood density to decrease their future maintenance costs and to enable greater reproduction. The common pattern for species often growing in gaps is to first grow rapidly thanks to low wood density and to later increase the density, while the opposite pattern is more common for species initially growing in shade.[24]

An ideal plantation species would be fast growing but would have decay-resistant, dense wood. Rapid growth is a pioneer trait, and both dense and decay-resistant wood are more typical for slow-growing nonpioneers. The lower maintenance costs associated with dense wood and the lower risk of microbe infection associated with wood with plenty of defensive compounds are both less valuable for fast-growing species exhibiting short life cycles. Therefore, foresters have tested and favoured species that are exceptional and that grow rapidly while still producing valuable timber such as *Tectona grandis*.

6

Tree Growth

Try to think like a tree. Obviously, trees do not have a nervous system and cannot actually think, but sometimes it can be useful to use our human terms and believe that trees want things. Animals have a conscious will, but much of their development and functioning is not dependent on what they want. A dog can decide whether to eat more but not how its digestion works. Exactly the latter is true for trees, and a tree's will is therefore very different from a dog's will. However, both wills have increased evolutionary success or fitness in the environment in which they have evolved. It is safe to assume that evolution has shaped all organisms in a direction that has resulted in greater fitness in the evolutionary past with the available evolutionary innovations. Therefore, the pet dog and its owner both eat too much because eating a lot has reduced the risk of starvation in the evolutionary past. In the same way, trees do wrong things in a managed environment. For example, a roadside tree surrounded by pavement might develop a thick bark protecting its trunk from the heat of forest fires. Therefore, to understand how trees think, we must consider their lives in the natural forest of their evolutionary past. So, what types of trees have been fit and have enabled successful reproduction?

Trees want to be as large (Weird thinking 6.1) as possible. In general, large size enables greater photosynthesis due to greater height and therefore higher light levels but also because of greater leaf area. Great height also enables wide wind dispersal of pollen and seeds and helps in avoiding a lethal temperature from surface fires and the mouths of herbivores foraging at ground level. Not all tree species reproduce only when large. In species-rich forest communities, many species, possibly even the majority, are understorey trees that photosynthesize and reproduce well in the shade of canopy species. Their life strategy of slowing down growth and focusing on reproduction already at a small size has been programmed into their genes. The proportion of small-stature tree species may be significant,[1] but their contribution to biomass[2] and photosynthesis[3] is typically dwarfed by canopy trees. For canopy trees, not reaching the canopy is a

Trees and Forests of the World. Markku Larjavaara, Oxford University Press. © Markku Larjavaara (2026).
DOI: 10.1093/9780197757109.003.0006

failure that prevents them from reproducing successfully. Therefore, in their evolutionary history, it has been critical to reach the canopy and to remain there as long as possible. Typically, trees are either killed, for example by diseases, or are overtopped by their neighbours. However, as trees adjust their risk-taking based on the competitive situation (Chapter 3), distinguishing these factors from the others is not possible. For example, an individual lagging behind in height growth may lower its energetically expensive defence and be killed by a pathogen. The rare individuals that make it to the canopy allocate significantly to reproduction only once they make it there. Normally, these winners continue reproducing for decades before senescence. Many slow species may even survive in the canopy for centuries. Sometimes these canopy trees stop or reduce their height growth and focus on reproduction. They may even sometimes be overtaken by their neighbours continuing their height growth. I have observed this pattern in individuals from the genus *Pinus* that have grown in open agricultural or urban settings, which stop their height growth well before their physiological maximum if height competition seems to be lacking initially and are unable to continue their height growth much later even if their neighbours surpass them in height.

Weird thinking 6.1 How to describe tree size?

A key statement in the main text is that trees want to be as large as possible. But what does "large" mean in this context? Is it trunk diameter, tree height, whole-tree biomass, or something else? One might first think that defining the measure of size is not important. Often this is the case, but choosing the size measure more carefully could potentially have changed or even reversed the study conclusions in a large body of scientific literature.

Most forest scientists use diameter at breast height or diameter at 1.3 m height as the measure of tree size. This is natural for us for two reasons. First, the lower parts of trunks are what we see when walking in a forest and what we can measure easily. Second, we, as animals, are not used to considering variation in density to influence size, as animal density does not vary much. By contrast, wood densities range nearly 20-fold.[20] Is then an individual of a light-wood species, with a slightly larger diameter at breast height, larger than its heavy-wood neighbour with about 20-fold biomass and strength in its trunk?

continued

continued

From the perspective of a forester selling timber priced by volume, the heavy-wood trunk really is smaller. However, in the context of this chapter, the heavy-wood individual is much larger. It can lift its crown much higher, or, alternatively, grow a much larger crown with dense foliage without compromising stability. Therefore, diameter at breast height is not always a suitable measure of size in an ecological context. Dangerously, using diameter at breast height in studies focusing on wood density inadvertently leads to comparisons of ecologically small light-wood trees to larger dense-wood species in cases when same diameter at breast height or basal area is the basis of comparison. Instead, in ecological studies with variable wood densities, whole-tree biomass, or trunk strength, estimated based on wood density (Chapter 5), would serve as a better basis of analysis.[21]

It is interesting to compare the timing of growth and the reproduction of plants and animals. Annual plants typically grow first and then begin accumulating energy for reproduction without growing much. Finally, just before the unfavourable season, they use all the accumulated energy to reproduce vigorously. Most birds and mammals are similar to annual plants and grow initially, then reach maturity and begin reproducing. Insects do not grow as adults because of their hard exoskeleton. Fish, reptiles, and perennial plants, including trees, are different, as they normally continue growing after the onset of reproduction. Intriguingly, whales, the giants of the animal world, are exceptions among mammals and are similar to the giants of the plant world because they continue growing their entire lives. The overly simplistic explanation for these patterns would be that being large is beneficial for these giants but for many other organisms, such as most mammals, there is an optimal size above which the functional capacity of the individual decreases. However, the picture is not this straightforward, as the energy allocated for further growth is away from reproduction, and even with larger optimal size it might be better to allocate the resources to reproduction rather than to growth.

All trees, when not considering palms, grow thickness and most grow height their entire lives; the answer to the question of why trees are the size that they are could simply be that they have not grown larger yet. To understand tree size from this perspective, we need to know the age and determinants of growth rate. Growth depends on both nutrients that trees obtain from the soil and carbohydrates coming from photosynthesis. Both tropical[4] and boreal forests[5] are strongly nutrient-limited; growth can be boosted by fertilizing the soil. Trees have various ways to deal with nutrient scarcity (Weird thinking 2.1). Trees are nearly always carbon or energy limited (Sceptic's question 6.1). As both nutrients and carbohydrates are required, thorough understanding of growth should be based on both, such as by incorporating them in detailed mathematical modelling.[6] However, soil nutrient status varies within biomes[5] and locally, and general patterns are hard to describe. On the contrary, energetic patterns are more suitable to be discussed globally, and therefore that is the perspective in this book.

Sceptic's question 6.1 Is growth limited by carbon?

Many researchers have advanced the idea that trees are not carbon limited. The element carbon is central for all life. In ecosystem science and ecophysiology, processes such as photosynthesis and decomposition are typically described as fluxes of carbon. This idea of trees not being carbon limited seems to originate from three areas of study.

Scientists have studied how trees respond to increasing atmospheric carbon dioxide content by releasing it on the upwind side of living trees. Carbon dioxide, together with water, is a raw material in photosynthesis and therefore an increase is expected to boost photosynthesis, GPP, and tree growth. The results from these studies have been mixed,[14] but typically the additional carbon dioxide has initially boosted growth, but this effect has faded away with more time since the initiation of the experiment.[15] The surprisingly low growth enhancement has led many to question the traditional thinking that carbon and carbohydrates are critical in tree growth and light competition. However, even if the bottleneck is not in the carbon uptake of leaves, it can still be further down the path of carbon from the leaves to its final uses. Carbon dioxide concentration

continued

continued

---

has not been as high as currently for millions of years,[16] and therefore it is hardly surprising that trees are unable to further benefit from even higher carbon dioxide contents in relation to what they have experienced in their evolutionary past. Analogously, we should not be surprised if photosynthesis is not boosted by light intensities two or three times greater than the maximum levels measured in full sunlight.

The second line of reasoning leading many to think that tree growth is not limited by carbon starts from datasets of nonstructural carbohydrate stocks that trees use as energy reserves. These reserves are normally plentiful,[17] which has led many to believe that as carbohydrates are continuously available for energy, they do not influence growth. This reasoning does not consider the very likely ability of trees to balance their growth based on reserve levels. Similarly, thinking that a person's income does not influence their consumption if their bank account is never empty would be flawed.

Based on the third perspective, photosynthesis, which is the source of carbon, might not be normally limiting, but instead, trees cannot use all the available carbohydrates. Clearly, some experimental data support this thinking and tropical seedlings, for example, grow faster with warmer night temperatures,[18] but these situations should probably be considered exceptions. In normal conditions growth is very much limited by carbon, and this is central to understanding tree evolution and their competition and physiology. The carbohydrate levels that trees can use determines how fast they can grow, how many seeds they can produce, and how small a risk they can take relative to mortality agents.

It is easier to describe where the carbohydrates are coming from than where they are going. Photosynthesis produces glucose, which is then converted into other carbohydrates. For a given tree species, assuming that all leaves receive approximately the same light levels and total photosynthesis per unit area, then GPP (gross primary productivity) would scale roughly linearly with leaf area. When imagining a monoculture plantation starting from bare ground, GPP does not increase as rapidly as expected from the leaf area. Only the top leaves are exposed to full light levels, but the leaves underneath are shaded by the upper leaves, reducing photosynthesis. All canopy species can use full light levels, but some do not tolerate any shade. Under a monoculture of light-demanding trees, plenty of light penetrates through the canopy and is available for other plants or possibly more shade-tolerant tree seedlings. Not all trees belong to the canopy species, and such species tolerate shade.

Trees use the carbohydrates that they produce in two basic ways. They are either used to construct structures, such as the trunk or leaves, or as energy to enable biochemical processes. These process functions are diverse and include anything from protection against pathogens to the transport of carbohydrates in the phloem downwards from the leaves. The carbon efflux out of the tree resulting from these processes is called autotrophic respiration. The carbohydrates used in constructing structures run out after senescence in heterotrophic respiration from microbes decomposing the material. From the tree's perspective, the timing of autotrophic respiration closely matches the timing of the benefit that the tree gets, but the same is not true for heterotrophic respiration as the structures are first part of a living tree. This conventional approach of dividing GPP into autotrophic respiration and net primary productivity of a material that is later decomposed, causing heterotrophic respiration, is practical because these concepts correspond to fluxes that are measurable in field studies. However, in successional forests which is what most of the world's forests are it is useful to make a distinction between the construction of short-lived tissue, such as leaves, and tissue that accumulates biomass, such as trunks.

The unusual approach that I have used, together with some colleagues, is to classify how the energy from photosynthates is used in maintaining a given level of biomass and what is then left available for accumulating biomass. This can be done at the scale of a single tree or for a hectare of forest. A large part of the autotrophic respiration of an individual tree is required even without the tree growing larger. This maintenance respiration, together with frequent replacement of the leaves and fine roots, form what I call "maintenance cost" (Weird thinking 6.2). The other part of carbohydrate use related to growth is partly autotrophic respiration related to building new structures and partly the carbohydrates ending up in the new structures. Because trees continue growing, at least during the growing season, this maintenance cost is only a theoretical concept that cannot be measured directly.

Weird thinking 6.2 What do construction and maintenance costs mean?

These are central concepts in this book. In economics, the magnitude of costs are easy to express with currencies. What is the currency for the construction and maintenance costs of trees?

Before considering the currency, let's look at various examples of construction and maintenance costs. In the simplest case, when trunk diameter is doubled, the amount of wood needed is four-fold and the construction cost of the wood is therefore also four-fold. When wood density varies, it can be assumed that the construction cost is still proportional with biomass. I have assumed that it is and that halving the diameter with four-fold wood density keeps the wood construction cost the same (Chapter 5). Things become more complicated when the constructions of differing tissue types are compared. For example, fats have a high ratio of energy to mass and are therefore useful to trees for keeping their seeds light weight yet filled with plenty of energy for the germinating seedling. For trees, synthesizing fats is certainly more costly than producing cellulose, but the cost ratio may not match the ratios in energy released in combustion and could depend on species and temperature, for example. In addition to carbohydrates, the construction of many tissue types also requires nutrients obtained from the soil (Weird thinking 2.1). Then the ratio of costs of constructing tissue with or without nitrogen depend on its scarcity in the soil and how the tree can obtain it. The additional cost of having nitrogen in the tissue is lower if the tree can obtain it easily. This could be the case when nitrogen is abundant in soil, if the tree is able to obtain atmospheric nitrogen with symbiotic bacteria, or if the species is exceptionally good in sucking up scarce soil nitrogen thanks to associated mycorrhiza, and is therefore associated with nitrogen-poor soils.

In this chapter with the ecosystem science perspective, part of the maintenance cost needed to keep biomass constant actually comes from the construction of leaves or fine roots or, with old-growth forests, even from the construction of entire trees to replace dead ones. However, part of the maintenance costs come from maintaining a stable structure, and energy originating from the photosynthesis is needed for that just as for construction.

It first seems hopeless to think of a common currency that would incorporate not only the costs paid with carbohydrates coming from leaves
and nutrients or even water from roots. However, as when pondering
whether interest rates could be used to describe urgency (Weird thinking 3.2), the linking can be done from an evolutionary perspective. Two
costs are of the same size if they reduce lifetime reproduction and fitness
equally, when not considering the benefits resulting from paying the cost.

I have defined the maintenance cost at a larger forest scale somewhat
differently: even though individual trees grow, if the biomass of the entire
stand does not increase, all energy goes into maintenance. In this case, the
maintenance cost corresponds better to empirically measurable fluxes as,
by definition, steady-state old-growth forests, such as many natural tropical
rainforests, allocate all their available energy into maintenance. At this scale,
maintenance also includes the replacement of tree individuals in addition to
the replacement of leaves and other short-lived tree parts.

This maintenance cost approach, in which both a tree's energy consumption and the building of structures needed to maintain a given biomass are
bundled together, is useful for understanding what happens in successional
forests. In an even-aged monoculture sown on bare ground, growth is initially very slow but increases nearly exponentially. This is possible because
at this stage, both GPP and maintenance costs scale almost linearly with
biomass, and therefore a given proportion of the available carbohydrates
can be used for growth. Only the self-shading of the lower leaves slows
this process, as explained above. In practice, this nearly exponential growth
means that the initial development of tiny seedlings after germination is
surprisingly slow, whereas the later growth of saplings with a height of a
couple of metres is surprisingly rapid to most observers. Subsequently, when
space is no longer overabundant, the canopy closes (Sceptic's question 6.2),
and a fierce competition for light really begins (Weird thinking 6.3). The
energetics change fundamentally. Because gravity, wind, and biomechanics do not differ greatly around the world, tree size at canopy closure does
not vary much, whereas age does, ranging around 10 years in the lowland
tropics and some 50 years in the boreal.[7] GPP cannot increase because the
leaf area per unit land area has reached its cap due to self-shading, but the
maintenance costs increase because when trunks enlarge, they also have

more energy-consuming sapwood. The stable GPP but increasing mainte-nance costs lead to the decreasing growth of individual trees and decreasing biomass accumulation rate at the stand scale. As sapwood maintenance uses very little energy compared to other tissue types, such as leaves,[8] the decreasing trend of biomass accumulation is slow but inevitable because building larger structures without more living tissue needing maintenance is impossible.

Sceptic's question 6.2 What happens when branches cannot extend further?

It is common to model tree structure by assuming that when trees become larger, their shape does not change or changes little based on a simple preset rule. Trees then become larger until they fill all the space avail-able. This has been used to model tree structure in general but has also been applied to specific research questions, such as those related to a self-thinning stand. Modelling involves simplifying the object or process that is modelled and therefore models always deviate from reality. However, a good model should deviate as little as possible from the truth with the chosen level of complexity (Sceptic's question 8.1). Is it realistic to assume the space-filling nature of trees in a stand?

When natural succession or growth of a plantation is observed, self-thinning does not begin for a long time after canopy closure. Instead, all trees typically continue growing, and they have more leaves in their crown top. However, their leaf area cannot grow without causing exces-sively shady conditions to their lowest leaves. Therefore, the lowest branches die out and the crown rises. This process can be assumed to be related to the lower branches touching each other and therefore being incapable of further growth. This is far better than assuming a given shape for the trees, but this thinking can be misleading as well.

Tree species vary tremendously in what kind of crowns they develop in a given competitive stand condition.[19] I once roughly gauged the crown of a huge *Samanea saman* tree in a park in Bogor, Indonesia, and esti-mated it to cover nearly a quarter of a hectare. Even though the crown of the largest *Ceiba pentendra* that I saw in Panama did not cover quite as much area, its branches spread more impressively above the main canopy of an old-growth forest. At the other extreme are many gymnosperm species (Chapter 4) that even when growing over 10 m tall in the open, have a crown less than 1 m wide. Sometimes even closely related subspecies, such as *Picea abies abies* and *Picea abies obovata*, have very different crown widths but similar GPPs and growth. Therefore, leaf area is a better basis for modelling compared to crown dimensions, unless the approach is species or subspecies specific.

Weird thinking 6.3 Do trees help each other?

A best-selling popular book describes how trees help each other.[22] The book portrays how *Fagus sylvatica* seed trees are mothering their seedlings to survive in the cruel world. In addition to huge popularity among the public, the book has been cited well among environmentalists. In reality, there is a long tradition of studies showing that seedlings further away do better than those under the protection of the mother tree,[23] as the negative impacts of shade, pests, and pathogens influence proximal seedlings more. Therefore, trees have evolved numerous techniques for dispersing their seeds far, partly to avoid these negative impacts but also to spread to new areas. These techniques include winged seeds flying far and animal dispersal. The popularity of the book perhaps tells more of human psychology than of trees. Many of us seem to want to classify not only people as good and bad guys but also other organisms in the same way, and trees are often perceived as good. As in the human world, good guys altruistically help others, especially small ones. Any evidence that supports good trees really being good is liked by many people as it backs their prior thinking.

In general, seedlings grow better away from other trees, but there are exceptions. Obviously, shade-demanding tree species need other trees to shelter them from the strongest sun and to moderate temperature fluctuations. Abrupt changes from nonforest to forest around the world indicate that no seedlings survive in the open, but that some survive in the forest. On the northern tree line, the slightly warmer microclimate within sparce stands may help the seedlings to establish and grow. In wetlands, such as boreal peatlands, large trees transpire and dry the soil so that seedlings have a deeper aerobic layer to uptake

continued

continued

---

nutrients. Within decades, tropical savannah can change into a dense forest because fires and browsing are weaker in the shady understorey (Chapter 3).

Favouring family members and kin selection is central in understanding the evolution of animals, including humans. The same evolutionary forces influence both animals and plants, and it is basically impossible that trees would not be influenced by how they are related to the neighbours, with which they compete for light and underground resources if they can detect them. There is clear indication that plant growth differs depending on whether their neighbours are their relatives or not.[24] I would expect trees to be more hostile towards unrelated neighbours and to use more allopathy and spread their roots and branches more aggressively towards unrelated individuals than towards their own offspring. In all, the motherly mother tree may truly support seedlings germinated from its seeds but based on quite different reasoning than that presented in the best-selling book,[22] and besides, the support is not normally strong enough to reverse the negative impacts of pests and pathogens.

Other explanations for the decreasing biomass accumulation and final stabilization have also been presented and tested. For example, challenges related to transport (Chapter 7) are likely to contribute to the slowdown of biomass accumulation,[9] but probably only slightly.[10] As discussed, most annual plants stop growing and allocate extra carbohydrates for reproduction in the latter part of their short lives, but trees instead continue growing. However, as the share spent on reproduction increases normally with age, this contributes to the slowdown in growth. The importance of reproduction in explaining growth can be seen from datasets with species showing great deal of variation in their reproductive investment between years.[11]

When energy is sufficient for maintenance but not for decent growth, the above-described growth slowdown in individual trees is a likely scenario. Alternatively, the subtle size differences that are always found even in even-aged monocultures favour larger trees while smaller trees die out, releasing light, nutrients, and water for the larger ones to proceed at a decent growth rate. This self-thinning law was described mathematically in the 1960s, and interestingly, thanks to the extra resources, larger trees become so much larger that the total biomass per unit land area exceeds that of the initial high-density stand.[12] Because a self-thinning forest is energetically close to the situation of an old-growth forest in which GPP is consumed for maintenance, the same principles can be used to model both maximal biomass and self-thinning. The lower maximal total volume or biomass of a dense stand is likely due to a large surface-to-volume ratio of thin trunks,[13] related to the high maintenance cost of woody structure surfaces (Weird thinking 5.1).

Most managed and nearly all natural forests are uneven aged and have multiple species of variable shade tolerances and maximum heights. For example, the understorey saplings of one species may grow well under the shade of a light-demanding species, while saplings of a less shade-tolerant species, that need all they can get from photosynthesis in maintenance, are stagnating and waiting for a lucky gap opening. These stagnating individuals are in a similar energetic balance as the giants of an old-growth forest and the slim trees in a plantation at the onset of self-thinning.

7

Tree Height, Diameter, and Energy

Why is a tall tree not one metre or one kilometre tall, but something in between? What limits tree height and how? To answer these questions, we need to consider the three main functions of tree trunks: storage, transport, and mechanical support. In addition, trunks serve the individual by photosynthesizing, but its significance is relatively small.[1]

Trunks store nonstructural carbohydrates for future energy use (Sceptic's question 6.1), water for dry periods, and also minerals.[2] Some tree species, such as *Adansonia digitata* or baobab, have unusually tapered and thick trunks, probably to increase the volume available for storage.[3] However, most species seem to have enough space in normal-sized sapwood to store the water and nonstructural carbohydrates that they need. If not, they could have evolved to have less heartwood or lower density, both of which increase the volume that can be used to store without additional biomass investment. Besides, the storage function of trunks cannot limit height, but instead taller trees have more space in their trunks. The other two functions, transport and support, conversely become more challenging the taller a tree grows.

Trees transport carbohydrates downwards in the phloem and water with solubles or sap upwards in the sapwood, which is also called the xylem. Transport and resulting inner pressure challenges in either the phloem[4] or the sapwood[5] can potentially limit tree height. Because the sap transport direction is upwards, challenges in this transport with increasing tree height can be caused by both path length and the elevational difference due to gravity, while downwards phloem transport problems can be related only to path length.

To discuss what limits tree height is bizarrely challenging, with two distinct semantic issues to consider. If something is restricting tree height below a limit but trees are far from this limit (Sceptic's question 7.1), can the factor be said to truly be limiting tree height? What if increasing tree height is impossible without causing changes to several independent mechanisms (Weird thinking 7.1)? Are all these mechanisms then limiting tree height, even if some evolve without difficulties when needed to

Trees and Forests of the World. Markku Larjavaara, Oxford University Press. © Markku Larjavaara (2026).
DOI: 10.1093/9780197757109.003.0007

allow greater heights? Several lines of evidence suggest that transport issues are not central problems that have restricted tree height the most in evolutionary history. First, lianas can be much longer than trees. They are difficult to measure, and field reports of exact length are lacking, but I remember a researcher telling me in Panama about a liana longer than a kilometre in his garden. As lianas face the same challenges related to path length as trees do, path length does not seem to limit tree height if other restrictions are lifted. Second, water transport structures are completely different in angiosperms and gymnosperms (Chapter 4); some angiosperms should be hundreds of meters tall based on the same physiological calculations that predict gymnosperm height realistically,[6] but their actual maximum heights are still similar.[7] Third, when considering global variation in tree heights (Chapter 8), the patterns are weakly explained by atmospheric vapour pressure deficit[8] or soil water availability within humid areas, which would be expected to play a major role if water transport was a central issue.

Sceptic's question 7.1   Does elastic buckling limit tree height?

If you transfer airplanes at Doha Airport, you have an excellent chance to familiarize yourself with the biomechanics of trees grown in windless conditions. Head for the central garden inside the airport, climb to one of the canopy bridges, grab a branch, pull vigorously, release, and watch the tree swaying back to its original position. Do not let the angry stares of the security guards break your concentration. If you have done something similar in a normal forest with normal winds, the scene at the airport is like a slow-motion film to you. These indoor trees are, similarly to those growing in the windless conditions of a rainforest understorey, close to the limit of elastic buckling.

When a self-standing column of a given diameter is elongated at some height, it will bend down due to gravity. Leonhard Euler developed the models of elastic buckling already in the eighteenth century, and interestingly, the height that leads to irreversible bending does not linearly depend on diameter. Instead, this maximum height scales to the diameter to the power of two-thirds[17]; that is, if diameter is doubled, then height can only increase by 59%. This relation has been explained to limit

continued

continued

tree height[18] and has been incorporated into many modelling approaches describing trees and forests.[19] In reality however, a significantly smaller diameter is sufficient to prevent elastic buckling at a given height and crown weight. A beaver must gnaw most trunk before the tree falls. Like beavers probably know, wind plays a major role, and this has been considered by computing a "safety factor" describing how far a given tree is from elastic buckling. The problem with this approach is that the extra trunk strength that is not needed to resist elastic buckling has very little to do with wind friction. Instead, bending moments caused by gravity and wind should be computed separately and then they could be compared to the strength of trunks at various heights. The actual taper of tree trunks closely matches the diameters needed to resist bending forces caused by extreme wind events,[12] and snapping is expected to occur with roughly equal probability at any height. However, unusual wind conditions or additional temporary weight, such as from sticky snow, may cause most trees to snap at a similar height.

It is a semantic question whether elastic buckling can be said to limit tree height. Undoubtedly, trees cannot grow beyond the height of elastic buckling, except if supported by other trees. However, other factors rather than elastic buckling are limiting most of the world's large trees, except exceptional trees such as those at Doha Airport.

Weird thinking 7.1 How to know what limits tree height?

An engineer friend of mine told me how he had once had no other option than to break into his own shed. The door looked fragile, so, hoping that he would break as little as possible, he grabbed the handle and pulled as hard as he could. The handle bent, the lock was damaged to a degree, and the door frame was partly loosened. He damaged everything but still could not get in. The door was designed perfectly. Just as his professor had taught him. None of the parts was overengineered. A badly designed door would break in one place, making the other parts too solidly and expensively designed.

Plants are designed both well and badly. Evolution is the most pedantic of all engineers, but when conditions are not spatially or temporally stable, the optimal design for a given set of conditions is far from optimal for another condition. Liebig's law of the minimum[20] describes how typically only one nutrient is limiting the growth of agricultural plants, and fertilization with other nutrients does not boost growth. In natural communities, plants are closer to the light–water–nutrient conditions in which they have evolved. The variation in vegetation between natural habitats is driven by species whose design has been optimized to variable conditions.

Water transport and biomechanical issues have been so constant in both time and space that trees have likely adapted perfectly to all of them. It is therefore acceptable to say that all of them limit tree height. However, their importance can vary. Similarly, as in perfectly engineered products, the cost of strengthening various parts may vary significantly, and it makes more sense to say that a given car has a given top speed because of its engine power rather than its tyre rubber quality.

To know which issues have been the most significant in limiting tree height during evolutionary history is challenging. One approach is to pay attention to current spatial variation (Chapter 8) and to draw conclusions from observations. Similarly, other variations in growth conditions can be used such as a comparison of free-standing trees and lianas. Or, physical barriers could be present that are not easily passed without new evolutionary innovations that we could try to identify. Often, the best approach is to use common sense to reason how fitness is influenced by the function under consideration. For example, significant mortality or the need to invest large carbohydrate levels reduce fitness.

The third function of tree trunks is to provide mechanical support. Unlike with storage and transport functions, the metabolically dead heartwood that has been expensive to build (Sceptic's question 5.3) also assists in biomechanics, suggesting that support is important. Furthermore, both trunk uprooting and snapping are common during storms and kill a large share of trees.[9] Therefore, wind damage has been clearly important for the evolutionary fitness of trees (Weird thinking 7.2). Mysteriously, trees rarely have branches that break, which could save the trunk. One possible explanation is that the pathogen load from the open cuts would overwhelm the defences and the tree surviving the storm would soon be killed by decay.

Weird thinking 7.2 Can you understand a tree trunk better than your own body?

All organisms adapt towards an optimal design using any available evolutionary innovations when conditions are stable. To understand what is going on and why a given trait is the way it is, we should understand trade-offs. What is the benefit and disadvantage when the trait changes towards one direction? Most trees that we see along streets and in plantations are far from natural forest conditions in which the species has evolved. However, we in our "human zoos"[21] are even further away from the conditions of our evolutionary history, and understanding the trade-offs in tree biomechanics may be easier than those related to our own bodies.

When thinking of trees with identical crowns, a tree with a larger trunk diameter is disadvantaged in the upward competition for light and a tree with a thinner trunk risks being snapped in a storm. Therefore, most trees have a nearly optimum diameter. Instead, we humans seem to be far from being optimally built if not exercising. This also holds when considering of societies of the distant past. Think of how useful a large muscular body would have been for a young man facing fights within his community to secure a high position in its hierarchies. Or a superb aerobic capacity allowing him to run after the last remaining individuals of megafauna and even escape if his initial attack was unsuccessful. However, larger muscles and higher aerobic capacity increase the energy consumption at rest,[22] which was harmful when food was scarce and instead fat was a better energy storage. Now our sedentary lifestyle is far from the conditions in which we have evolved, and life expectancy increases both from aerobic exercise and from training our muscles.

The health benefits of exercise can be divided into those related to the energy consumed while exercising and those resulting from processes preparing our bodies for similar challenges in the future. The latter process is parallel with the acclimation of trees. Being barely able to lift a weight sends a signal that more powerful muscles are needed for challenges ahead even though the energy consumption at rest increases.

Similarly, the vigorous bending of a trunk signals that a thicker trunk is needed, or the next storm might prove fatal.[23] Sport scientists understand these processes from their empirical datasets, but we forest scientists still lack basic information on how seasonality and flexing intensity influence trunk diameter growth. However, as from an evolutionary perspective the setting is analogous in both cases, it seems likely that an extreme bending almost snapping the trunk is what triggers rapid changes while moderate flexing may not influence growth at all. Analogously, moderate exercise may benefit human health only by increasing the energy consumption of that particular day, but does not improve our fitness for future endeavours.

The evidence suggests that biomechanical support is the key function of tree trunks. As both trunk uprooting and snapping occur in the same forests (Chapter 5), it is reasonable to assume that the strength of the root system and associated soil is roughly in balance with the strength of the trunk, as both coarse roots and the trunk principally serve to keep a tree erect, and extra construction costs for overengineering one over the other would be wasted. We can thus focus on trunks and have confidence that roots are providing similar strength.

Have you tried to snap a shrub or a small tree of approximately your own height simply by pushing the top down towards the ground? If so, you know that the stem does not crack easily and once you release it, it bounces back, sways back and forth a few times, and then looks exactly like it did before the whole procedure. The question arises as to why trees cannot similarly develop trunks that can bend all the way to the ground and then bounce back when the gust of wind is over. This shrub biomechanical strategy[10] (Chapter 2) is possible with two conditions. First, the stem has to be able to bend to the ground. When a stem is bending, fibres within the wood stretch on the convex side and compress on the concave side. The further away they are from the middle or neutral axis, which does not experience lengthening or shortening, the more they bend. With a given bending, doubling the diameter doubles the strain of the outermost fibres. It does not matter much whether we consider diameter at the commonly used breast height or 1.3 m or another height closer to the base of the trunk. Because the fibres can resist only a given amount of strain without breaking, woody plants need to be thin relative to their height to be able to bend to the ground. However, thin woody plants may not be able to support their own weight because of elastic buckling (Sceptic's question 7.1). The ratio of diameter needed to resist elastic buckling relative to a plant's height increases with increasing size. As the maximum diameter of being able to bend to the ground scales linearly with height but the minimum diameter to resist elastic buckling does not, woody plants following the shrub strategy inevitably hit a brick wall and cannot grow taller.

Larger trees are different to shrubs, but their biomechanics are nevertheless often modelled misleadingly based on elastic buckling (Sceptic's question 7.1). In reality, wind is often more important than gravity. Trunks bend away from the wind, and fibres are stretched on the windward convex side and correspondingly compressed on the other side (Chapter 5). When trunk diameter is doubled, the cross-section area is four-fold, and the number of fibres is also four-fold. However, on average, the fibres are then double the distance from the neutral axis with a four-fold resistance to bending per fibre. Four times more fibres, with each offering four-fold resistance to bending, results in a whopping 16-fold resistance against bending.

Imagine a narrow river that is crossed by a road bridge that sways a few metres from left to right on windy days. Even if the critical joints would not fail on the first day, the road surface would deteriorate rapidly, and even before that few drivers would have the courage to drive on it. Engineers normally want to design rigid structures (Chapter 5). Now imagine a tall tree swaying a few metres on a windy day. It looks completely normal, as significant swaying is typical especially for tall and thin trees. Whether more or less bending is advantageous is unclear and can depend on the tree (Chapter 5). The more bending occurs, the further to the side the centre of gravity is and the more gravity contributes to the bending moments of the tree trunk, which could snap due to this extra force. However, trees may survive the strongest gusts thanks to their flexibility if the gust is very short, if the sail area of the crown is streamlined enough, or if the crown bends down to a height that is sheltered by neighbouring trees.

Bending is important for understanding elastic buckling and the shrub strategy, but whether bending has been good or bad in the evolutionary history of trees remains unclear. However, trees obviously do not want their trunks to snap. Because fibres can resist only a given amount of strain and because the outermost fibres stretch or compress more with larger diameter and with a given bending, the maximum bending of thicker trunks is smaller. Doubling the diameter leads to 16-fold resistance to bending. However, as the maximum bending is only half of this, the strength or the maximal moment that the trunk resists bending is eight-fold. Even though eight-fold resistance to bending is less than 16-fold, it is still more than the four-fold material investment when doubling the diameter. Therefore, trees typically concentrate their trunk building into a single stem (Sceptic's question 7.2).

Sceptic's question 7.2 What are the pros and cons of having multiple stems?

A couple of years ago, I spent a few weeks doing precommercial thinning. My idea was to leave the most vigorous 2,000 trees per hectare to grow. On several occasions on a *Pinus sylvestris*-dominated hill, I made the wrong choice but luckily looked upwards just before sawing. *Pinus sylvestris* is the main winter fodder for moose in Finland, and even though the trunk bases looked good, the trunks forked at the height of moose browsing, a couple of metres above the ground, and had shorter and wider crowns.

The scientific literature on multistemmed trees forking at the base is affected by scientists comparing single-stemmed and multistemmed individuals of the same thickness. This is misleading, as the multi-stemmed individual is biologically larger (Weird thinking 6.1), making this a comparison between a smaller single-stemmed tree and a larger multistemmed tree. A more fruitful starting point is to compare individuals with the same biomass. A single stem with a diameter of 40 cm has the same biomass as four stems of 20 cm because volume scales with the square of the diameter, assuming the same height, taper, and wood density. However, the strength of these stems scales with the cube of the diameter and, assuming their strength is needed only to withstand the force of the wind in the leaves, the single-stemmed tree can be double the height assuming the same wind speed further up or, alternatively, can have double the number of leaves. The superiority of having only a single stem seems even more overwhelming when considering the lower surface-to-volume ratio (Chapter 5). However, having a single stem is not solely advantageous. Shrubs or young trees with a biomechanical strat-egy of being able to bend to the ground unharmed cannot carry much leaf mass on a single stem and therefore may need several stems. Multi-stemmed trees have an insurance of sorts that lowers their mortality risk,

continued

continued

---

as a falling larger tree is unlikely to kill all the stems. Finally, branches of a given length reach further when their points of attachment are already some distance apart, which is useful for spreading the crown. This is particularly useful for species with a more aggressive strategy towards their neighbours. The high stem surface area is perhaps less harmful and the possibility of relying on the shrub biomechanical strategy more useful for fast tree species, while the insurance against total destruction is more useful for slow species.

Multistemmed trees often shorten their life cycle and grow more in width and less in height, possibly because of more wood needed to support a similar tall crown. Alternatively, the wound from an injury may cause decay to spread in the wood, or the tree can be prepared for potential decay shortening the length of the optimal life cycle.

I criticize basing our understanding of tree structure on gravity acting only on trees mass alone without considering wind (Sceptic's question 7.1) and potentially other loads such as water, ice, or snow attached on tree surfaces. Ideally all of these should be included, and sometimes most of them are.[11] However, I argue that if gravity or wind has to be excluded for simplicity, we are better off focusing on wind alone. Several case studies indicate that this is the better option.[12] When additionally assuming that the bending moment is caused by wind friction solely in the leaves and not in the branches and trunk and that leaf area per land area is constant, as it is typically very stable in monocultures after canopy closure.[13] Then we can relatively easily estimate how the trunk diameter that is sufficient for resisting the bending over moments changes with changing stand characteristics.

With a given stand density (i.e. the number of trees per unit area), and therefore wind friction in the crown of one tree in an even-aged monoculture, doubling the trunk diameter leads to eight-fold strength, which is then able to resist an eight-fold bending moment. As the wind friction for a tree is fixed, the tree can be eight-fold in height. Things are somewhat more complicated when stand density is allowed to vary. For example, still assuming an invariable leaf area per land area, thinning of seven eights of the trees, and allowing acclimation and crowns to occupy the space available results in each tree having eight-fold leaf area, eight-fold wind friction, and therefore requires double the previous diameter for a given height. In practice, however, both diameter and height are increasing in a growing stand.

I began this chapter by asking why trees are not a kilometre tall. So far, I have explained that they need to have a certain diameter to be able to resist the bending moments from wind. The above reasoning is therefore not sufficient for understanding size per se, but is useful for understanding the height–diameter relationship. Based on these biomechanics, height is not limited if diameter is not. To be precise, columnar structures that are very wide relative to their heights can collapse downwards without bending, but the structure has to be much wider than that of a typical tree. In practice, something unrelated to mechanics has to also limit tree size.

I previously described how, in an ageing stand with larger and larger trunks, more and more of the energy in the form of carbohydrates, mostly sugars, end up being used just to maintain the biomass (Chapter 6). This energetic perspective can be applied in a larger ecosystem scale in which replacement of entire tree individuals is considered a maintenance cost required to keep biomass at a given level or at a small-individual-tree scale. The latter is not entirely accurate, because most trees cannot stop growing and more than just the maintenance cost is needed to stay alive. Nevertheless, consider a stagnating tree that uses all its energy just to maintain is a useful approximation. Then, as the focus was on an even-aged monoculture with an invariable leaf area per unit land area, the maintenance cost of roots, branches, and leaves can be assumed invariable, and the maintenance cost imposed by trunks depends on sapwood volume and trunk surface area. As sapwood cross-section area is proportional to leaf area,[14] with a given leaf area its volume is therefore proportional with tree height. Trees have sufficient energy to maintain themselves at a given size only when the energy consumption from tree height and trunk surface area is small enough.

Tree size is restricted by both the energetic and biomechanical limit. When these are compared in the same graph, we see how the larger diameter allows greater height biomechanically, but by contrast energetically trunks should be thin to be able to be tall (Figure 7.1). I computed the biomechanical limit with the same assumptions described above so that doubling

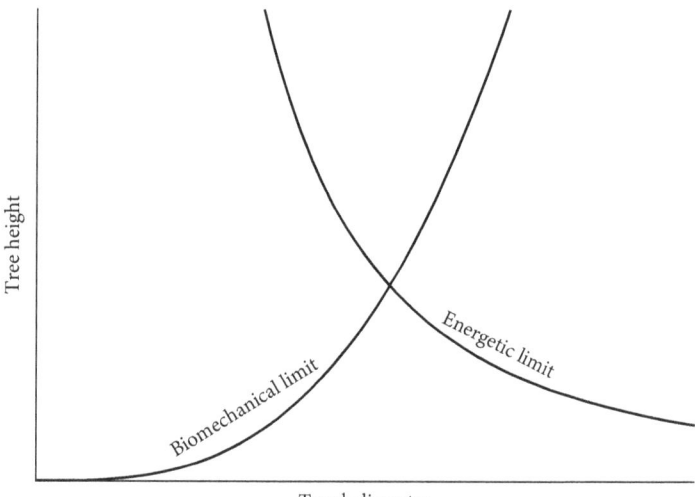

Figure 7.1 Biomechanical and energetic limit of tree height with a given leaf area. Because trees try to be as tall as they can be without crossing the limiting curves, their trunk dimensions approach the intersection of the two curves at a given stand density from below but always towards the right as diameters increase.

diameter allows height to be eight-fold. The energetic limit is based on assuming that surface area causes all trunk-related maintenance costs. If sapwood volume caused all the maintenance costs, the energetic limit would be a flat line. More realistically, when both incur a cost, the limit is a less steep curve.

Even though this view (Figure 7.1) is a gross simplification, it has been very useful for the nearly 20 years that I have walked in forests around the world. The situation is obviously far more complex in natural forests with dozens of species of all sizes, with understorey trees experiencing lower wind speeds and biomechanically less demanding circumstances but lower light levels and energetically more challenging conditions.[15] However, envisioning a simple even-aged monoculture can be a useful starting point for understanding how changing conditions impact trees. Such understanding could later be applied to more complex forest ecosystems. For example, changing wind speeds alter the biomechanical limit. Stronger winds shorten the height that is possible with a given risk. Changing risk-taking impacts both curves. If a tree is taking a higher risk related to winds, the biomechanical limit is raised. Correspondingly, increased risk-taking related to energy stored for bad times lifts the energetic limit and increases tree height. This strategy can backfire during difficult times such as a drought. Pioneers are expected to take higher risks related to both limits, to benefit from being taller, yet have higher mortality. The mysterious self-thinning law (Chapter 6) can be modelled simply by computing the intersections of the two limits for variable stand densities.[16] If photosynthesis is boosted, such as by increasing soil fertility, the energetic limit is pushed upwards. Water availability and air temperature have similar impacts, and those are discussed next (Chapter 8).

8

Climate and Size

One of the greatest hikes of my life was in a national park where the world's tallest tree lived. I remember the giant *Sequoia sempervirens* trees along the trail rising out from a green carpet of ferns like massive pillars of ancient Greek temples. Camping was allowed only close to a beach, where the *Picea sitchensis* trees looked small and unimpressive. In the morning, with more time having passed since gazing and getting used to the possibly 100-m tall *Sequoia sempervirens* trees, the perhaps 50-m tall *Picea* trees seemed much larger than the previous evening. The tallest *Sequoia sempervirens* in Redwood National Park in northern California is 116 m tall.[1] I have been lucky to see many other tall tree species. The tallest *Eucalyptus regnans*, which I saw in Victoria Australia, was probably over 80 m, and the tallest *Sequoiadendron giganteum* in the Californian mountains was likely over 90 m. In North Cascades National Park, I carried a rangefinder and knew that I was sleeping under a leaning 81-m *Pseudotsuga menziesii*. In various parts of Malaysia, I have stared at towering trees mostly belonging to the family Dipterocarpaceae; in the mountains of southern Chile I saw huge *Araucaria araucana* trees; and, most recently, on the southeastern slopes of the Himalaya in China not far from Myanmar, I saw nearly as large *Abies georgei* individuals, even though the tallest ones did not end up in our research plots.[2] Even though I have not travelled everywhere, I have seen the largest trees of most regions. This has not been pure luck, but more truthfully, my unforgettable experience in Redwood National Park in 2002 made me want to see more and led me to wonder why the tallest trees grow where they grow.

In his book,[3] Diamond argued that Eurasian and North African people have agriculturally and technologically developed the most advanced cultures, as the land mass were large. This allows adapting an innovation, such as the domestication of an animal species, over a large area. Therefore, a given location benefits from a greater number of innovations. A similar process certainly influences tree heights. When thinking of a small volcanic island that has recently emerged from the ocean, few would expect

Trees and Forests of the World. Markku Larjavaara, Oxford University Press. © Markku Larjavaara (2026).
DOI: 10.1093/9780197757109.003.0008

the first tree species arriving there to break the world record for height, for several reasons. First, the recent arrivals may not be perfectly adapted to the new locations. Second, their numbers are low and the tallest species may be missing from the first arrivals. Third, the most successful species in recently forested regions may be fast species, while slow species evolve over millions of years. Based on this thinking, it is not just the age and size of the island that should influence the height of its tallest trees, but also the distance to larger land masses both currently and historically. Oceanic islands, such as New Caledonia, that were previously connected to larger land masses would be expected to have taller native trees than islands with no historical connections such as Hawaii. There seems to be support for this theory among small islands but not on continents, where the central parts should have more potential species migrating to them. Instead, the tallest tree species seem to mysteriously thrive on continental edges.

If most of the global variation in tree height cannot be explained with past and current connectedness of the land masses, could climate be the determining factor? If you ask people how the redwoods can grow so tall, many will explain that the key is the summer fog on the northern coast of California. Fog can help in two ways. First, water may drip down to the ground and be sucked up from the soil by trees like rainwater. This has been shown to be important for *Sequoia sempervirens* trees[4] because it can help them during the summer months with very little precipitation. Second, trees may uptake water directly from the surfaces of their leaves as *Sequoia sempervirens* trees do.[5] This has the advantage that water does not to have to be transported a long way up from the roots to the crown, but the disadvantage is that this shortcut makes sucking soil nutrients impossible. Fog helping the tallest trees in the world is a nice story, but can it explain where the tallest trees grow? The dripping fog water is just a partial substitute for the lacking summer rain, and areas with abundant precipitation should be even better for tall trees. Direct foliar water uptake may have some importance for one of the world's tall species but is unlikely to have a more global impact.

If fog is not important, what other climatic variables could play a more significant role? Only some tropical trees develop annual tree rings. Therefore, to study interannual variation in diameter growth, our project installed stainless steel dendrometers around the trunks of trees on Barro Colorado Island, Panama. I was surprised when a field technician showed me a dendrometer that had snapped by melting in a thunderstorm while the tree looked completely unharmed. Global lightning densities suggest that lightning is a significant ecological factor in many biomes and therefore trees must have adaptions to survive it, like the lucky one in Panama, which probably had the current running on the surface of its bark and not in the wood. Indeed, based on rigorous research, lighting is a significant killer in tropical forests.[6] One way to lower the risk of being hit by lightning is to be less tall relative to neighbouring trees. With very frequent lightning, this could, in theory, lead to trees becoming shorter and eventually becoming shrubs to stay safe. There is no evidence of this, and the regions with the most frequent lightning in Southeast Asia and South and Central America[7] naturally have dense forests with tall trees. However, it seems plausible that emergent trees are influenced by lightning, as they may lower their risk of being hit by growing width instead of height, which has implications on light competition and on the prevalence of angiosperms and gymnosperms that seem to be handicapped in spreading their crowns (Chapter 4).

As the case of lightning avoidance shows, maximum heights and biomasses do not correspond perfectly but certain pointy gymnosperms may be dozens of metres taller than angiosperms with same biomass but wider crowns. For example, even though the tallest tropical tree, a 99-m tall *Shorea faguetiana* angiosperm in Borneo, Malaysia,[8] is shorter than Asia's tallest tree, a gymnosperm from the genus *Cupressus* at 102 m in China close to the easternmost tip of India,[9] the wider crowns of the angiosperms may complicate the size contest, and the slightly taller gymnosperm should possibly be considered biologically smaller (Weird thinking 6.1). However, trees have large trunks to have a large leaf area high up, and this correlates well with tree height even though some variation occurs. Similarly, there is a strong link between individual tree sizes and biomasses per unit of land area.[10] Therefore, the questions concerning the distribution of tallest trees and greatest old-growth forest biomasses are closely related.

As I described previously, when trees become larger, all carbohydrates are used for maintenance, as larger trunks take up more of the relatively invariable GPP (Chapter 6), ultimately reaching a steady-state old-growth stage. If GPP is larger, the old-growth biomasses should increase. GPP is largest in lowland tropical rainforests. These forests have huge biomasses but not the largest in the world.[11] GPP drops with decreasing temperatures and precipitation. Trees may not be able to photosynthesize fully to avoid escaping water while opening their stomata. Low precipitation can also influence maximal biomass indirectly via increasing fires or even more indirectly by favouring tree species that have a fast life cycle because of stand-replacing fires (Chapter 3). However, increasing precipitation in humid climates does not increase canopy heights,[12] and thinking based solely on GPP and precipitation is not sufficient to explain why maximal biomass varies in intermediate and high-biomass forests of the world.

When focusing on humid forests only, we can ignore precipitation and concentrate solely on temperatures. The energy budgets of trees depend not only on GPP but also on maintenance costs that are comprised of autotrophic respiration resulting from living tissue metabolism; replacement of tree parts, such as leaves; and, in an ecosystem-scale examination, also the replacement of dead trees. Clumping all these processes into one concept is a gross simplification, but simplifications can be useful (Sceptic's question 8.1). We can assume that maintenance costs increase with temperature while the temperature dependency of GPP is very different (Figure 8.1). GPP is basically absent in freezing air temperatures and increases first with increasing temperature. However, this pattern reverses due to multiple physiological reasons and finally collapses at very high temperatures that are not found in current humid climates of the world.

Simultaneously examining GPP and maintenance cost curves in the same graph (Figure 8.1) is useful for understanding climate-caused variation in maximal biomass. Greater biomass incurs a larger maintenance cost, but the

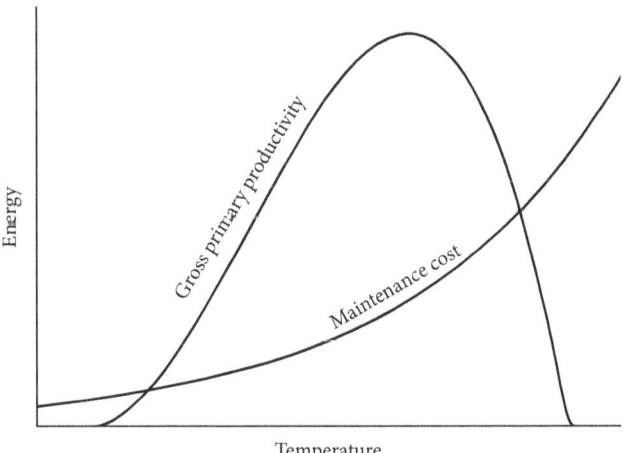

Figure 8.1 Dependence of GPP (gross primary productivity) and the maintenance cost of air temperature. The maximal biomass peaks in climates with the highest GPP relative to maintenance costs, while biomass accumulation in young forests is fastest in warmer temperatures close to the GPP peak.

temperature dependency of the curve remains the same. Assuming a constant temperature, maximal biomass is attained at the temperature with the greatest ratio of GPP to maintenance cost. This graph can also be used to understand the biomass accumulation rate in a successional forest. When GPP is high relative to the maintenance cost from the successional biomass, the biomass accumulation rate is determined by the difference between the two curves. When the forest is young and biomass is small but the canopy has already closed and GPP has reached its maximal level (Chapter 6), the maintenance cost curve is at a low level and biomass accumulation is peaking close but slightly to the left of the GPP curve peak.

Sceptic's question 8.1 What is the optimal complexity level when modelling forests?

While writing I must constantly consider how much technical language I use. More readers understand what I mean when I use simple structures and words, but these take up more space and slow down the communication for masters of the specific terminology. Unfortunately, many experts attempt to exclude alternative thinkers from discussions (Sceptic's question 12.1) or enhance their scientific authority by using overly complex language. Some also favour vague words such as "sustainability", to make analytical critique difficult (preface).

Statistical or empirical modelling summarizes datasets facilitating generalizations and is used in nearly all empirical forest science. This then allows, for example, testing whether trends or differences are statistically significant and allows other scientists to base their research on these simplifications. Techniques to balance simplicity and realism in statistical modelling are available and widely used,[20] but again, especially early-career forest researchers seem to inappropriately think that more complex equated to more scientific and thus to more valued.

When forest scientists talk about modellers or when I mention the word "model" in this book, the focus is on process-based or mechanistic models in which processes and their relationships are described mathematically.[21] This understanding can then be used to understand certain processes based on other, better-known processes. In a way, a complex process-based model is a theory described mathematically instead of using words. When the temperature dependency of GPP and maintenance cost (Figure 8.1) are linked mathematically to something else, such as maximal biomass, this could be considered an extremely simple process-based model. Unlike in scientific language or in statistical modelling, the typical thinking has been that the more detail used to describe processes the better. If outputs from the model have not corresponded with reality, more processes have been included and details have been added to the processes already described. This is certainly much better than tuning the parameters of the underlying processes, but unfortunately it readily leads to overly complex models.

Dynamic global vegetation models are as complex as the name suggests and include thousands of processes between vegetation and their environment. In his talk at the Carbon Dioxide Conference in 2017, Colin Prentice described how the success of climatological models used for weather forecasting encouraged plant modellers to attempt something similar. However, he continued explaining that these vegetation models have not really been that useful. Even though the models get the observed patterns right, as this has been ensured during their development, they disagree on the mechanisms,[22] divert considerably from newly available global datasets,[23] and produce drastically differing predictions for the future.[24] Because of this variation, clearly not all these models can be right and possibly all of them are wrong.

Models of various complexity levels are needed in forest and vegetation science, but it is good to advance slowly and listen to dissidents raising their voices during the process. Describing some processes in detail does not help much when others are weakly expressed.

We parameterized the models of GPP and maintenance cost with a dataset of old-growth forest biomasses from around the world.[13] The ratio of GPP and maintenance cost peaks around 15–18°C, suggesting that the tallest trees and highest biomasses should be found in humid climates with temperatures closer to this range. Instead, fastest growth is expected in warmer temperatures around 21–24°C for youngest forests.[14] This difference in optimal temperatures for maximal biomass and fast growth has implications on the optimal distribution of the world's production and carbon storage forests (Chapter 14).

Based on the normal approach to understanding growth (Chapter 6), the analogous procedure to explain maximal biomass would be based on net primary productivity and turnover. The higher the net primary productivity and the lower the turnover, the higher the maximal biomass. In both approaches variation in tree species or tree community position in the fast–slow continuum influences both productivity and turnover or maintenance cost. Similar maximal biomasses result in a given climate and soil from high GPP and maintenance cost or low GPP and maintenance cost. Correspondingly in the normal approach, net primary productivity trades with turnover causing unexplained variation in attempts to parameterize the models with empirical data. Even though the maximal biomasses are similar, actually slow species communities often have somewhat higher maximal biomasses (Weird thinking 2.1) but are trailing behind in recovery rate after disturbance (Chapter 3).

In general, seasonal temperature variation shifts the warm and cold seasons away from the optimal temperature for maximal biomass or fast biomass accumulation, but there are exceptions (Weird thinking 8.1). Seasonal temperature variation is smallest close to the equator in the tropics and in the most maritime climates elsewhere in which the high heat capacity of ocean water warms up the winter and cools down the summer. Humid lowland tropics have mean temperatures of around 26–27°C, which is too warm for the world's largest trees to support themselves energetically. Instead, optimal temperatures for highest biomasses are found on the most exposed parts of continents or higher up on mountain slopes in the tropics. When thinking of the world's temperate maritime regions on continents with suitable climates,[15] most have very large trees, with Western Europe, from Scotland and Denmark south to northernmost Spain, being a notable exception. However, the suitability of the Western European climate is supported by the very large Australian and North American trees growing in this area. Strangely, native, tall, and high-biomass species seem to be lacking in Europe, but potentially this has something to do with the depauperate tree species richness.[16]

When considering forests at a higher elevation above sea level, the picture is messier. Some tropical or subtropical mountains have extremely tall trees. The tallest trees of Africa[17] and Asia[9] grow at a significant elevation, as does the huge tropical lowland species living at almost 500 m above sea level in Borneo.[8] However, field ecologists in some studies have reported decreasing forest biomass upwards from close to sea level on a humid elevational gradient.[18] Even though energetically temperatures should be more suitable for extreme size at higher elevations, sites high up may have shallower soils, reducing soil fertility, water, and consequently GPP, or they may be more prone to landslides, and local species are therefore adapted to a short life cycle, or they may have depauperate species communities because of climatic isolation. For extreme tallness, in addition to perfect climate, these species should also have a narrow crown and a very slow life cycle, such as the *Sequoia sempervirens*, which continues growing height even at the age of a thousand years.[19]

Weird thinking 8.1 Do trees benefit from variation?

It is not weird thinking but rather common sense that if an optimal exists, then deviations from it are bad for trees. For example, when considering how air temperatures influence maximal biomass, a good first guess is that there is an optimum and it is better that the temperature stays within this optimum without any seasonal or interannual variation. A deviation from it would be harmful for two reasons. First, temperature would then be suboptimal, and second, any change may be problematic for a tree acclimated to the previous temperature. If the second mechanism is important relative to the first, a constant suboptimal temperature to which the tree has acclimated might be better than experiencing a change towards the optimal. However, it is more complicated than this. As maximal biomass can be understood based on GPP and maintenance cost having very different dependencies on temperature (Figure 8.1), any temporal variation in another factor influencing GPP, such as sun elevation angle, may cause temperature variation to become useful. As with the case of maximal biomass, trees should benefit from colder temperatures during nights and from winters with longer nights and lower sun elevation angles at daytime. When the mean temperature is very low for maximal biomass, the seasonal temperature variation, even when not considering sun elevation angle or other complications, can be beneficial because at least the boreal summer is suitable for trees instead of the permanently unsuitable tundra climate.

Interestingly, theoretically similar patterns exist with precipitation. Abundant precipitation seems best to be distributed evenly, instead of flooding and anaerobic soils during the wet season and drought conditions during the dry season. However, similarly as in cold climates,

continued

continued

it is probably also better that the rains in very dry climates come during a clear rainy season, so that the trees can evolve their phenology to follow a clear cycle and so that the abundant water recharges ground water reserves that are more utilizable by deep-rooted trees rather than shallow-rooted smaller plants.

Similar patterns are recognizable far from the temporal variation of climatic variables. Intensively cultivated fruit trees might have close to perfect soil fertility and acidity, which leads to maximum yield, such as the optimal temperature or precipitation to maximal biomass. If you work in such a fruit orchard, you better fertilize the soil just as recommended based on a meticulous soil fertility analysis. However, if you are a hobby gardener, soil fertility and acidity may be far from the optimum, and it might be better to be sloppier and to cause spatial variation in the soil. This variation could be even more helpful than the seasonality of temperatures in cold climates, as fruit trees could obtain one nutrient from one direction and another from the opposite bearing.

PART III
FORESTS BENEFIT PEOPLE

9

The History of Forest Use

Imagine a world without humans. Consider the vegetation, especially the trees. What do you see? Whatever your view is, it is likely to be far off from real virgin-state nature. It is very difficult for us, who have never seen such forests, to first consider all the mechanisms through which humans are currently impacting nature, and even more difficult to understand what would have resulted if these mechanisms had not happened over the past millennia.

Lightning ignites natural fires, but these are rare in many biomes[1] because thunderstorms need exceptionally humid air to develop during the dry seasons when fuels are driest. Even a sparse human population can drastically influence ecosystems if they either handle fires carelessly, causing large unintentional conflagrations, or by intentionally igniting large fires when fuels are abundant and dry because they consider large blazes to be beneficial, such as to increase the number of huntable or gatherable products. Even the very earliest humans may have benefitted from the opening of ecosystems and from deforestation (Sceptic's question 9.1). These human-caused fires may have significantly increased fire frequency, potentially causing drastic changes in ecosystems shifting from closed forests to savannahs due to positive feedback, as a burned area may reburn more easily (Chapter 3).

Sceptic's question 9.1 Did humans evolve from apes thanks to deforestation?

In the late 1980s, as a young teenager, I began reading popular science books not about trees and forests, but about human evolution and evolutionary psychology. Only a few had been translated into Finnish, but they were good ones, and these books taught me to think from an evolutionary perspective. One of the theories that I remember from those years is the "savannah hypothesis", which I recall stating that the drying of tropical African climates some millions of years ago led to deforestation and

continued

Trees and Forests of the World. Markku Larjavaara, Oxford University Press. © Markku Larjavaara (2026).
DOI: 10.1093/9780197757109.003.0009

continued

to the apes adapted to forests experiencing a need to change, resulting in bipedalism that freed their hands to use tools and led to more carnivorous diets and complex social dynamics. If this is true, then deforestation and nonforests would have played a central role in early human development, to the horror of some of us who are forest lovers.

Most of the steps in this reasoning appear to be supported by modern science. The fossils of early hominids have indeed been formed in locations that had an arid savannah climate at the time.[24] Also, rapid evolution in a novel environment is a reasonable assumption and is commonly described in empirical studies of other animals. However, the thinking that a climatic change caused the initial push is unlikely true.

Major changes in human prehistory, such as biological evolution, cultural evolution, or migrations, have often been explained by rapid climatic changes. This is not just the case in popular science literature. We have been told that people lived happily and harmoniously with their environment (Sceptic's question 9.2), but then a negative change forced them to adjust. This thinking seems unreasonable for two reasons. First, the urgency to obtain more food was the norm and was not restricted to abnormal times, including climatically challenging. Second, climatic changes were much slower than the current anthropogenic climate change, and therefore biomes shifted slow relative to human dispersal ability. Therefore, even without a climatic change, early hominids were able to walk out of the forests and expose themselves to new conditions, potentially triggering rapid evolution.

In addition to fires, hunting seed disperses and browsers or other animals that influence them is a second way in which a sparce human population can drastically modify ecosystems at a large scale. Again, the picture is hazy. Prehistoric hunting pressure has varied little in space and time in many regions, and envisioning nature without human hunting is therefore difficult. However, Australia and the Americas were populated by humans only some tens of thousands of years ago, and the simultaneous extinction of megafauna unadapted to coping with the new bipedal social hunters points to the importance of hunting already in prehistoric times,[2] even though it might be discomforting to think about primitive people influencing their environment (Sceptic's question 9.2). Similarly, recent sudden changes in hunting or comparisons between adjacent areas with differing hunting regimes highlight the importance of hunting in how humans influence ecosystems. As a consequence of prehistoric extinctions, many tree species that are still common today but are decreasing in abundance have fruits that are not dispersed by any extant animals but were rather eaten and spread by species of the extinct megafauna.[3] Some attempts have been made to explain more recent hunting and seed dispersal impacts or potential impacts,[4] but these are riddled with excessive simplifications.[5] By contrast, the hunting of large herbivores can have direct and very significant impacts on vegetation.[6]

Sceptic's question 9.2   Did prehistoric people live in harmony with nature?

Idealistically, we like to think that prehistoric people wanted to take care of their environment and lived in harmony with surrounding nature. As legislative tools and economic incentives were not feasible, moral codes on what not to do and primitive systems of sanctions were in place. For example, when I was living in rural Burkina Faso, a friend told me, in the middle of the dry season, that deciduous *Adansonia digitata* or baobab trees are holy and should not be damaged. At the onset of the rainy season when food is scarcest, I understood the rational of this taboo when people began eating sauces made from flushing baobab leaves. Similar taboos tracing from the prehistoric past are common around the world, but they nevertheless only cover a tiny fraction of all natural resource use. The Malthusian divergence between the six children that were born and the two that were possible to keep alive made prehistoric people grab nearly all they could eat. The evidence of early human impacts show not only animal extinctions but also carbon budgets, demonstrating early changes.[25]

Similarly, we like to think that Indigenous people are saviours of trees. Indigenous people are "good guys," and trees are also liked (Weird thinking 6.3), so people are naturally excited when reading how Indigenous people want the best for trees. The term "Indigenous people" is clear and useful in the Americas and Australia, to which the majority of the ancestors of the current population have migrated from far during the past two

continued

continued

---

or three centuries, but difficult in Africa and Eurasia, in which the waves of migration have been less distinct. For example, Finnish people have been living in southern Finland for a millennium, after the Sami people. The Sami were the dominant population for perhaps a millennium before the Finns, although the full picture is complex, with languages and genes often moving asynchronously. Finns are not considered Indigenous, as we are the dominant population in the nation, but were we Indigenous before our independence from Russia in 1917, when we were a minority in the context of the Russian empire? Ignoring the tricky term "Indigenous," does the family and cultural connection to those who have used the land centuries ago influence how sustainably the environment is used? There is certainly more understanding of the matter. I not only learned about the techniques of swidden cultivation, in which my grandfather took part as a child in the 1910s, but, more valuably, I also learned something of how swidden farmers think. But this connection does not seem to bring a harmonious balance with nature. Stagnating culture without technological development forced prehistoric and early Indigenous people to live relatively harmoniously with nature, but this was not due to their will to do so but rather to their inability to obtain more food for their starving children.

Into a third group I classify all the modern and intensive ways in which humans have converted land uses and modified ecosystems after fires and hunting. Their magnitude is probably easier to understand than those of fires and hunting, as we see these mechanisms in action around us: agriculture, silviculture, pastoralism, mining, transport, and urbanization.

The fourth group is the human-driven spread of organisms over oceans. It has already tremendously impacted the fauna and flora of small islands with depauperate species pools to begin with, but the changes on continents have concentrated in human-disturbed ecosystems. However, this does not mean that wider impacts are not happening, as the productivity of more natural forested ecosystems is very low relative to the biomass, changes are slow, and it could take millennia before exotic old-growth tree species spread significantly.

Finally, the fifth group includes other indirect global change impacts in which humans first influence the composition of the atmosphere which then impacts the world's ecosystems. I discuss the effects of climate change, carbon dioxide, and nitrogen fertilizations later (Chapter 15).

Of the five above-mentioned mechanisms regarding how humans have shaped the environment, fire has probably been used for a million years.[7] Hunting-driven changes have occurred for much longer, but their impacts have unlikely been as significant as the effects of fire. The third mechanism group, the more direct land-use impacts, is much more recent. The domestication of mammals, such as sheep, goats, and cattle, over 10,000 years ago[8] made killing them much easier; enabled the regular harvesting of their products, such as milk and wool; and permitted breeding desirable traits. These browsers and grazers grew well in similar Poaceae-dominated nonforests as well as their wild and hunted predecessors and many other large herbivores, and people had similar interests for using fire and other means to kill trees to ensure the openness of the ecosystems. However, as the production system of human food became simpler, more predictable, and more productive, enabling a larger population density, the efforts to open up the landscape intensified. However, agriculture became much more significant than animal husbandry in terms of the energetic value of the produced food.

Astonishingly, agriculture developed independently at least twice. The timing was about the same as for animal domestication. It led to land uses that are considerably more different from natural nonforests than pastures are. Once again, fire has been used both in the conversion process from other land-use types and in subsequent cropland maintenance, but less than in rangeland management in which fire was used in large areas annually. The domestication of new plant species and breeding of cultivars thriving in various soils and climates led to the area of cropland increasing over time. Finally, after the migration of large numbers of Europeans to the Americas and Australia 100 to 300 years ago, most globally suitable land for pasture or cropland was in use. Unsuitable areas consisted of exceptionally steep slopes and unfertile or waterlogged soils, and on a larger scale, of land in climates with too short a growing season due to aridity or low temperatures or too humid climates preventing the use of fire in the conversion process (Weird thinking 9.1). This led to the disappearance of large continental-scale forest areas outside of the equatorial tropics and boreal climates, which have poor soils, a growing season lasting only a few months, and intermittent frosts that may kill crops. For modern-day people, it may be difficult to comprehend how laborious the conversion of a lowland rainforest into a managed non-forest has been prior to fossil fuel-powered machinery developed in recent decades. Simply cutting down a single, large, dense-wooded tree may have taken days using hand tools, and a more permanent conversion would have required burning and killing forest tree seeds, sprouts, and seedlings for weeks and months after clearing. Even after this, if the dry season is not dry enough during some years to dry the felled trees and ignite a high-intensity fire, the hard work may have been wasted. The diseases of humans and domesticated animals have been another challenge in the humid tropics.

Weird thinking 9.1 How to deforest without a bulldozer?

Yes, you can also use an excavator. But how to do it without any motorized equipment—not even a chainsaw? In this case, trees are a huge problem for you. One way to clear land for agriculture is not to deforest, but to find an area without any trees growing naturally. In most tree-less areas, the soils or climate are not conducive for plant growth, but there are exceptions. Seasonally flooded plains may have anaerobic soils for part of the year but can be extremely fertile at other times, and can be suitable for short-lived plants such as rice.[26] The challenges of keeping other plants away are not restricted to trees or to the initial clearing, because weeds also need to be kept away or at least weak permanently. Again, flooding can help. I once observed, in Indonesia, an off-season extremely dry paddy suddenly inundated and ploughed before rice was planted. The uprooted weeds adapted to aridity had no way of competing in growth while submerged with the planted rice.

Clearing a high-biomass forest for agriculture without motor-powered equipment is difficult, and close to impossible without using fire. For the first years, trees are sprouting vigorously from the stumps and roots; thereafter, the weed control challenges are similar to paddy rice cultivation but without the possibility of flooding. In Laos, while watching a recently burnt swidden used for producing so-called upland or dry rice without flooding, I asked about the yield compared to paddy rice and was surprised to learn that the yield is similar but the workload to keep the weeds away is manyfold, highlighting how the flooding is central for weed management. When modern machinery, flooding, or fire cannot be used, farmers still need to create a sudden change in conditions to boost the competitive advantage of the crops relative to the weeds. Then, the best option is uprooting the weeds with ploughing or tilling, often using draught animals, and sowing or planting immediately after. Or, alternatively, the agricultural system is more based on trees rather than smaller plants, as in the pre-Columbian Amazon basin.[27] It is much easier to keep trees thriving than small crops when they are competing with small or medium-sized weeds (Sceptic's question 13.1).

The reality of life with a steady population density and nonexistent or primitive contraception is another aspect that may be difficult for modern-day people to comprehend. For example, if a woman gave birth to six children,[9] assuming a steady state, only an average two survived into their reproductive phase and the rest died, often due to diseases after being weakened by malnutrition. It is not difficult for modern-day people even in wealthy countries to comprehend how offspring weakening due to malnutrition caused a tremendous concern, and parents did their best to obtain more food and farmers cleared more cropland if that was possible. The more food was available, the more children survived. However, as Thomas Malthus argued,[10] improvements in food production only led to a temporary relief, as the increase in population density meant that even more food had to be produced to end the misery. Because the fertility rate greatly exceeded two, only very exceptionally did people have sufficient food for consecutive decades. These exceptions perhaps occurred only briefly after migrations to uninhabited lands or to areas inhabited by people belonging to cultures that were less efficient at producing food in those climates and soils. Similarly, the rapid technological and scientific advancements in the late nineteenth century lowered the risk of starvation in industrialized countries. Naturally, wars and other violence, pathogens, and predators also limited the human population size, but food production was clearly the main factor and human population density principally depended on climate, soils, and food-producing culture and technology.

Finally, in the twentieth century, widespread contraception brought fertility rates down.[9] Agricultural productivity concurrently increased rapidly thanks to manufactured nitrogen fertilizers, efficient breeding techniques, and combustion engine-powered machinery.[11] These together lowered the mortality from starvation close to zero towards the end of the twentieth century, except in the world's poorest countries located mainly in Africa or during wars or other calamities. Perhaps surprisingly, the increased productivity thanks to technological advancement and rapid slowdown in the increase of mouths to feed, and the simultaneous shift from self-subsistence agriculture to producing for markets and associated increased productivity, did not initially lead to a decrease in managed nonforest area. Instead, faster deforestation ensued with better machinery, thanks to which even tropical rainforests could be cleared, supported by modern treatment of tropical diseases in humans and domesticated animals. Vastly greater crops were produced, but an increasing share went to domestic animals as fodder.

The rapid increase in agricultural crop production was followed in many regions by migration to urban centres. This occurred because despite the agricultural productivity increase, industrial productivity had increased even more rapidly and pulled labour towards more lucrative jobs. Agricultural workforce scarcity led to even further mechanization, increasing the productivity per person, but also caused a lack of profitability in conditions like stony soils or steep slopes where mechanization was challenging. The subsequent developments depended on governmental policies. If subsidies boosted the prices of agricultural products, intensive agriculture continued. Instead, farmers were incentivized to lower production intensity when subsidies were based on cropland area. Commonly, the establishment of plantations has been subsidized, resulting in rapid forestation. Without significant agricultural or forestry subsidies, the least valuable cropland or pastures were simply abandoned, normally resulting in natural forest succession. Active or passive forestation reversed the forest area trend and has been called forest transition.[12] Interestingly, the same processes are repeated in countries at a similar economic developmental level in all biomes and continents, with little difference in whether the population is indigenous, recent immigrants, or something in between (Sceptic's question 9.2).

The above-described trends have been dominated by croplands and food production. However, many other processes related to the products and services that ecosystems can provide to humans have also influenced forest area trends. Timber harvesting has potentially been the most significant service, as the product was first used as fuel and then to build huts, houses, and vehicles, and finally to make paper, cardboard, and veneer. To explain the continuum of various harvesting regimes, I focus on the two extremes that I call sustainable and unsustainable harvesting. Unsustainable harvesting has, from an economic perspective, more in common with mineral mining than with sustainable timber harvesting. When unsustainable harvesting has been permitted, it has often occurred immediately after becoming profitable. This pattern could be called the "tragedy of the busy", as it has led to only modest revenues instead of potential manyfold profits after some decades of infrastructure development reducing the transportation costs.

Unsustainably harvested forests can have various trajectories after a harvesting event. Many are converted to pasture or allowed to regenerate back to a forest, but if valuable tree species have been harvested, the resulting forest may not enable profitable reharvesting for a century or even longer. Overall, the media, at least in Europe and the English-speaking world, seems to overemphasize the importance of industrial logging as a cause of deforestation. In the tropics, for example, where agriculture plays a major role,[13] logging has been a significant player only in tropical Asia with short distances to cities and the coast and where a single tree family, Dipterocarpaceae, has dominated the species assemblage, with most species being valuable for industries.[14] Instead, in Africa and the neotropics, unsustainable logging has typically focused on a smaller proportion of species,[15] causing minor impacts to carbon dynamics, biodiversity, and hydrology. In South America, large-scale agriculture has been the main deforestation driver. In Africa, subsistence agriculture has continued as the norm, as it was in much of the rest of the world prior to the last century. Overall, globally, nearly all deforestation this century has been linked to agriculture.[16] Sustainable forest management practices are discussed more from a silvicultural perspective in the next chapter, but socioeconomically plantations are typically closer to croplands than to natural forests. Only the rotation period is much longer, and the management is less labour-intensive.

When Jared Diamond began writing his book on societal collapses,[17] he assumed deforestation to be the main cause. However, during the research process he found that the causes were more complex in most cases, but deforestation still plays a part in most of the included collapses. In some locations deforestation influenced hydrology, leading to detrimental erosion. Therefore, it could be argued that the leadership of these societies was incompetent or short-sighted and did not place enough weight on long-term concerns. However, this may be undue criticism, as most land in most societies centuries ago was controlled by families or small groups of people mainly focusing on producing food. Only exceptionally did the central governments of larger countries have a strong influence on land use. Many Eurasian leaders reserved forested landscapes for hunting and other recreation, but this was not done for the benefit of the people but was rather motivated by more selfish aims.[18] However, already centuries ago, training the first generations of foresters, especially in French- and German-speaking parts of Europe, was motivated by concerns for the sufficiency of timber resources, but the results of these early attempts to save the forests were probably modest.[19] In contrast to the leaders and their advisors, common people often valued nonforests and the food they produced more than forests.[19] Japan probably experienced the first massive reforestation event, as the reforestation of Japanese mountain slopes happened as early as the early eighteenth century, but this was exceptional.[17]

Interest rates are most commonly used in forest sciences when discussing forest management (Chapter 10), but I use the same concept to describe the differences between fast and slow tree species (Weird thinking 3.2). Because forest development takes decades or centuries, even small differences in interest rate cause huge differences in whether an investment in a new forest is beneficial or not (Figure 9.1). Forests are favoured in many ways when optimizing with small interest rates. First, unsustainable logging stereotypically brings short-term benefits but may lead to dramatic problems related to erosion, hydrology, or timber supply in the long term. Second, planting trees in plantations or assisting natural regeneration is only optimal under low interest rate optimization. The investment of planting seedlings is often profitable only under low interest rates, as the future timber harvest and other products or services are available only after a long waiting period, which amplifies the role of the interest rate used. What then determines whether societies and their leaders make decisions that are optimal under low interest rates and favour forests? These societies and their governance are stable, there is no urgency or emergency, and decision-makers need to understand the technically more complicated benefits of forests relative to nonforests. For example, the Japanese government that organized the massive reforestation in the early eighteenth century was extremely stable.[17] Similarly, Chinese[20] and Vietnamese[21] regimes that have organized globally exceptionally rapid and massive reforestations in the past decades have also been exceptionally stable. The influence of the democracy level on forest area trends has attracted some research interest (Weird thinking 9.2), and it seems that the main impacts operate via the stability of the regimes.[22] Therefore, similarly as unstable dictatorships fear the next coup, the planning horizon may become very short in the current social media-driven ultra-democracies. Finally, it is worth noting that with a high interest rate and large-scale destruction of old forests and absence of new forests can be

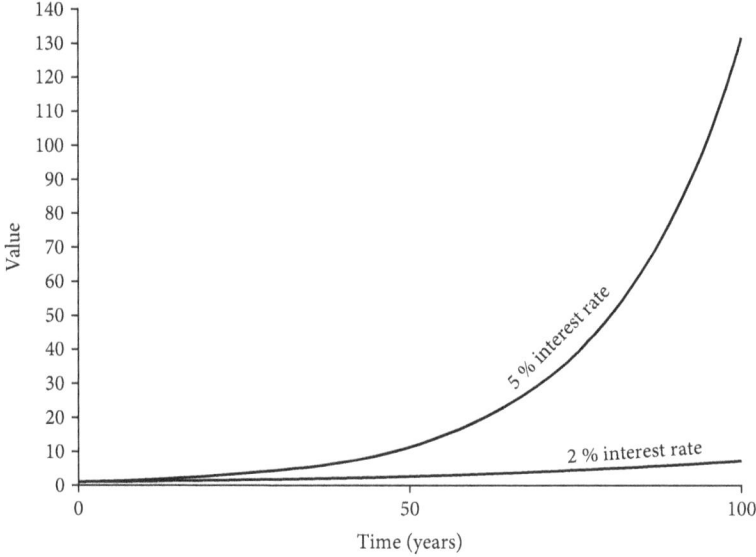

Figure 9.1 The influence of interest rate on whether an investment is profitable. Investments with an initial value of 1 are profitable if the value is above the 5% line and have at least an 11-fold value in 50 years and a 132-fold value in 100 years. The 2% line, with a 2% increase annually, climbs to three-fold at 50 years and to sevenfold at 100 years. Note that even though the difference between the lines seems extreme, many environmental economists argue for an interest rate close to 0% and industrial investors may expect interest rates well beyond 10% in their computations.

optimal in some situations. Analogously, it may be logical and wise for a family to borrow money under a high interest rate due to temporary sickness of the main breadwinner, for example, as it may be optimal to conduct unsustainable logging even if the area becomes wasteland, if there is exceptional urgency for the funds obtained from the logging.

Weird thinking 9.2 Does democracy boost forest area?

I was working for an international organization around 2006 and visiting forestation plots around southeast Asia. Many people talked about the problems related to democratic policies and practices, yet Western

continued

continued

governments and NGOs (nongovernmental organizations) were nevertheless pushing for bottom-up community forestry and participatory planning. This piqued my curiosity; I plotted various democracy indices and forest area changes and observed that poor democratic countries were indeed losing their forests while nondemocratic ones increased their forest area.[22]

The interesting part began after an official at the Finnish Ministry of Foreign Affairs read my scientific article.[22] I had thought that discussing the pros and cons of democracy and forest area would have been an interesting academic question, but to this individual it was a taboo topic, as democracy seemed to be a non-negotiable sacred value to him. During the following weeks, he then tried to cause me trouble by sending emails to the director of the Finnish Forest Research Institute where I was working at the time, asking whether my article represented the Institute's view. As free speech is often listed as one of the fundamental pillars of democracy, this fanatic proponent *of* democracy was paradoxically working against democracy when trying to limit analytical discussion *on* democracy.

The Western democratic world has changed in the past 15 years, as social media has enabled individuals to rise into influential positions without being part of established political parties or media houses. This process is nowadays widely criticized, but this criticism still does not consider that the systems are too democratic, as the use of the word in Western media is restricted to positive contexts only. Instead, the word "populism" is used for excessive democracy. Nonetheless, Western countries are trying to force other countries to democratize and follow other fundamental Western values, such as human rights. This cultural imperialism is often contradictorily combined with apologies for past Western cultural imperialism and colonialism. Respect for multiculturalism is signalled at the superficial level with an interest in exotic dishes and music.

It is not difficult to think of many mechanisms through which democracy either increases or decreases forest area. Democracies enable environmental NGOs and free press, and their leaders may interact more with scientists and other forestry experts. However, possibly more importantly, democracies may have problems in supporting long-term projects with no foreseeable benefits within the same election cycle, especially if the impact pathways are complex and difficult to comprehend for an average voter. Therefore, it seems that nondemocratic systems are more likely to boost forest area. If large forest area is considered positive, then nondemocracies may be better at achieving this, but their forests may not benefit people. Besides, democracy can be valuable for something far more important than forest cover percentage. It helps humankind avoid wars.[28]

Above, both natural forests and plantations have been favoured in low interest rate optimization. However, when land tenure is considered, natural forests and plantations are very different. When certainty of being able to control a piece of land is high, long-term investments may be profitable. However, when controlling a piece of land is uncertain, long-term investments, such as plantation establishment, are not encouraged. Therefore, increasing uncertainty increases the interest rate needed for the investment to be profitable. Similarly, uncertainty encourages rapid and unsustainable logging of natural forests, as the potential revenue may go to someone else. This situation is common in natural resource utilization and has been coined the "tragedy of the commons".[23] When the main driver of deforestation is conversion to cropland, the situation is very different relative to logging-driven deforestation. Uncertainty may protect forests from agriculture-driven deforestation because the work put into opening the ecosystem and working the soil to make agriculture possible may potentially be wasted soon. Such situations have probably been very common or even the norm in the past millennia. Therefore, weak governance and high interest rates can both save and destroy forests depending on the drivers of deforestation.

10

Wood Production

Have you ever heard of a university with a Faculty of Shrubland Science? Me neither, but I have visited many Faculties of Forest Science. In a given climate, shrublands are not far behind forests in leaf area or GPP, yet they do not provide timber, which is the perfect material for many types of construction. When building a house or a bridge, the optimal construction material should resist bending moments and decay, as the heartwood of living trees does (Chapter 5). Securing and maximizing the production of this valuable material has been a central question in the forest sciences, especially before the past decades with an increasing interest in climate change mitigation (Chapter 11) and some other benefits (Chapter 12). The terms "forest management" and "silviculture" have been used to describe this most deep-rooted area within the forest sciences. These terms can be used nearly interchangeably, even though typically "silviculture" has been used more narrowly, with less focus on economics and social matters and more focus on field techniques and optimizing the development of a forest stand for wood production. Forest mensuration (calculating how much harvestable wood a stand has) and forest technology, focusing on how to transport logs out of the forest, are other traditional subdisciplines within the forest sciences.

Let's begin with a simple case. An even-aged and even-sized monoculture plantation is typically planted from seedlings grown in a nursery. Alternatively, the seeds can be directly sown on the site or perhaps the sprouts from stumps or roots of the previous tree generation will begin a new one. Biomass accumulation is initially slow, then faster, after which it slows down again (Chapter 6). Harvesting timing, which determines the rotation period, can simply be chosen based on maximal biomass accumulation (Figure 10.1), but economically the optimal period can be far from this. Positive interest rates shorten the optimal rotation period, but more importantly, larger trees normally have a larger proportion of biomass in their trunks and larger trunk logs are more valuable per unit volume, postponing the optimal harvesting age. The value of logs with increasing size is caused by two separate mechanisms. First, the cost of harvesting and

Trees and Forests of the World. Markku Larjavaara, Oxford University Press. © Markku Larjavaara (2026).
DOI: 10.1093/9780197757109.003.0010

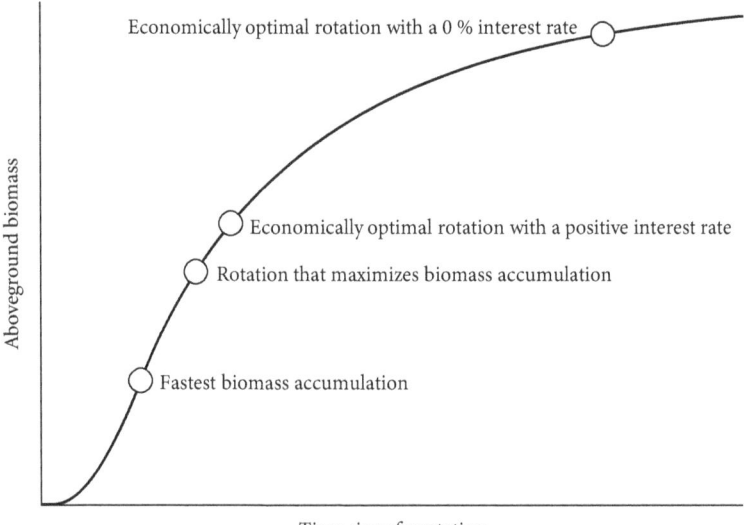

Figure 10.1 Biomass accumulation after forestation. The lowest circle shows the age at which biomass accumulation rate peaks, which corresponds to the age of canopy closure. The second lowest circle demonstrates the age at which a final felling maximizes the biomass production per unit time during the whole rotation. However, because larger trees have a larger share of biomass in wood, and the larger the tree, the more valuable the wood, the economically optimal rotation period is much longer, as shown in the topmost circle. Positive interest rates increase the urgency to obtain revenue from the wood and to shorten the rotation period, as shown in the second highest circle. The two top datapoints are hypothetical, but the curve and the lowest two circles show the above-ground biomasses of 44 and 83 mg per hectare at ages of 14 and 22 years, respectively, based on a global parameterization representing biomass accumulation for the first century for the climate of Wuhan, China.[4]

transporting per unit volume of wood decreases with increasing tree size. Second, wood from larger logs is of better quality, with a smaller proportion of wood close to the surface that is difficult to process into lumber, a higher proportion of decay-resistant heartwood, fewer branch knots, and the possibility of cutting large pieces of lumber. In addition, the planting or other costs at the beginning of the cycle similarly alter the optimal rotation period. When other products than wood are considered, the optimal rotation period can be shortened or prolonged. For example, many game species and harvestable berries and mushrooms favour young plantations while others favour old ones (Chapter 12). The most important services, such as biodiversity conservation, carbon storage, and hydrological benefits, normally become more valuable with increasing tree age and size, and therefore the optimal rotation period lengthens if these are considered.[1] Similarly, the harm caused by a sudden drastic change in the landscape may lengthen the optimal rotation, or even lead to some form of continuous cover forestry. In practice, however, wood production dominates rotation period optimization, and rotations range from less than 10 years in the lowland tropics[2] and small-diameter trunks for pulp manufacturing to approximately 100 years in boreal plantations, with the objective of producing logs large enough to be processed into lumber.[3]

There are hundreds of important plantation species for wood production around the world. Good guidebooks already exist that introduce these for a given country, such as for Panama,[5] and I therefore provide a more general and theoretical discussion of plantation species traits. I argue that it is important to make a distinction between traits that have benefited trees in nature in their evolutionary history and traits that are beneficial to humans but where a clear linkage to fitness in natural forests is missing. Obviously, a good plantation tree species grows rapidly. However, as fast growth has been important for all species in their evolutionary history, similar but slower growers have faced extinction. Therefore, focusing on these other traits that influence growth speed may explain the suitability of species for plantations. Trees seem to range in the level at which they focus on competing with their neighbours in terms of trunk growth or other means. Some species block their neighbours' growth by wide crowns and large leaves (Chapter 4). This pattern is unwanted, because from a forester's perspective, carbohydrates are wasted when allocated into growth away from the trunk, blocking off neighbours lowers their GPP and growth, and trunk wood quality may be reduced due to large knots in branches. Similar to the above-ground, tree roots can aggressively head towards their neighbours, but these patterns remain poorly known due to methodological challenges.[6] More extreme below-ground warfare may involve allelopathic chemicals[7] that are harmful to the aims of forest managers with wood production objectives if the chemicals inhibit the growth of neighbouring trees, but beneficial if only understorey plants are harmed and the energetic or nutritional cost of synthesizing the chemicals is not too high. Trunk growth can also be unusually slow for other reasons than aggressive competition towards neighbours. Species that allocate plenty of resources to early reproduction may be competitive in the natural setting, as the great number of seedlings may compensate for the slower growth. However, for a forest manager, plentiful seedlings well before harvesting age is normally just a waste, although these seedlings develop into subsequent production trees in certain silvicultural systems. Similarly, sprouts from the stumps or roots of harvested trees may be a valuable beginning for a new rotation.

Another area to consider when wondering whether a given tree species would be suitable for plantations is the continuum from fast species, which are often pioneers and which I refer to as short-sighted species, to slow species, which have evolved to resists disturbances and to reproduce late (Chapter 3). The rapid growth of the fast species is great for the forester, but it often tapers off at a relatively small size as the trees begin allocating more resources on reproduction. Even worse for the forester, these fast species typically take greater risks and have a higher probability of being damaged or killed per unit time. This is partly caused by their lower investment in defensive compounds in wood, which influences not only their mortality while still growing but also the utilization of the harvested timber. Furthermore, their wood is typically of low density (Chapter 5) and therefore more elastic and prone to ruptures and thus less valuable for most purposes. The lower wood density of fast species implies that the "rapid growth" is less rapid when the unit of size is biomass per unit area rather than the more usual trunk volume.[8]

Interestingly, the risk of a plantation tree being damaged or killed during its growth can be altered when species are introduced into new areas. For example, the risk of wind damage can differ in various regions, and not necessarily so that windier regions have more windthrows. As trees adjust the diameter and therefore the strength of their trunks to the wind climate, the wind speeds in extreme storms relative to winds during normal windy days may be what matters (Sceptic's question 3.1). Biotic disturbances like pests and pathogens may have evolved with their host tree species, and many tree species have been proposed to grow better as exotics, with less pressure from pests and pathogens than in their native range.[9] For similar reasons, exotics are often claimed to be much worse than natives when conserving local biodiversity.[10]

To simplify, if the rapid height growth of trunks is caused by columnar crowns, then this characteristic is beneficial for plantations. However, if the rapid growth is caused by a pioneer trait and high risk-taking, then the wood will likely be of low value, but the rapid growth may compensate for the cheap price of a unit volume harvested. However, there are several traits that have probably not been of primary importance in the evolutionary history of most species, but are important for utilizing harvested logs. A straight trunk, typical for all gymnosperms and many angiosperms of lowland tropical rainforests, makes transportation and processing simpler and results in higher-quality lumber. Most species can be grown to have a straight single-stemmed trunk, but the easiest species for the forester are those that develop in this manner even when grown sparsely. With the most challenging species, such as *Quercus robur* from the family Fagaceae, famous for their wide crowns,[11] even after a branchless trunk has developed, high density is required to discourage epicormic shoot growth. A large proportion of decay-resistant heartwood is another valued characteristic of a good timber species (Chapter 5). However, living trees have a trade-off between passive defence in the heartwood, requiring high initial investment against microbes, and active defence in the sapwood (Sceptic's question 5.3). Finally, specific traits, such as growth stresses, influence the ease of harvesting, and many other traits impact the workability of the wood. Several important plantation species, such as *Tectona grandis*, are well known for the ease with which the wood can be processed into various products. These same traits, which probably have not been of primary importance in evolutionary history, are more easily improved by breeding than growth speed is (Sceptic's question 10.1).

Sceptic's question 10.1 Is it easy to breed trees to grow faster?

Domestic animals have been bred rapidly to produce much more food, and pets have been bred to look cuter. By contrast, despite the huge incentives and efforts, long-distance racehorses still run at almost the same speed as they did decades ago.[15] Are trees more like racehorses, which have been difficult to breed to run faster, or like funny-looking dogs with flat faces that have been bred in just a few decades from healthy landraces to the point of almost being incapable of breathing?

Again, it is useful to think of the evolutionary past of trees and the traits that have been crucial for their fitness. All the ancestors during the perhaps one million generations of current gymnosperms have been successful, most of them in a closed forest, and have managed this by growing rapidly; by avoiding being killed by wind, animals, or pathogens; and by reproducing effectively. The resulting trees are already good at factors that have increased their fitness in their evolutionary past, and rapid growth is very much one of those traits. Boosting tree growth by breeding seems difficult and restricted to a zero-sum game in which a potential gain is a setback somewhere else that has been important in evolutionary history. To understand the options, we can think of the variation between species and the fast–slow continuum (Chapter 3): fast growth is traded with high mortality. Varieties accumulating trunk volumes rapidly could have lower density wood, and be more susceptible to windthrow, fires, pests, or pathogens. Or, alternatively, they could reproduce less.[16] If the forest management system is based on planting without natural regeneration, the lower investment in reproduction could be okay but the higher mortality would not be. Breeding typically narrows genetic diversity, which may be another mechanism increasing mortality, as biotic agents such as insects and microbes can spread more easily in genetically uniform populations.

Breeding has more potential if the aim is to use the fast-growing varieties in conditions that differ significantly from those of the evolutionary past. For example, breeding trees to grow faster in a climate that differs from that of their evolutionary past or shade trees to benefit from abundant light should be easier. Similarly, breeding a trait, such as branch thickness, that does not have the clear gradient from bad to good, like tree growth does, is expected to be rapid. However, any breeding with trees is slower than with herbaceous crop species that have a much shorter generation time.

Starting the development of a new tree generation is often the most unpredictable phase. Sprouting (i.e. coppice forestry) can be used in plantations of certain species, such as in the genus *Eucalyptus*, thus making initial management easier because sprouts typically grow faster than competing weed species. The sprouts of many species are not used in forestry for industrial wood, as the resulting stem may not develop into a large trunk and can also be infected from the cut (Sceptic's question 7.2). However, the lower-quality wood may not be an issue in firewood production, which still represents half of all the wood harvested.[12] Many species, such as most gymnosperms, do not sprout.

Perhaps we can first consider whether the rapid establishment would be guaranteed if the species is selected to match the soil and climate, the soil is prepared with scarification, the timing of planting for perfectly raised nursery seedlings is well chosen, and the young stand is weeded and perhaps even fertilized and watered. However, the first months and years after planting are risky. Many potential dangers, such as drought, flooding, pathogens, pests, or larger herbivores, can first weaken and then kill the young planted trees, and the site can then be occupied by other, unwanted tree species. It may be puzzling how the perfectly selected and well-cared-for seedlings still die and the site becomes occupied by a species that has perhaps been unsuccessfully planted at another location. An analogy from economic systems may shed light on this (Weird thinking 10.1). A natural succession or plantation that fails to develop as intended may be stuck for years or even decades because the seedlings are suppressed by species in the family Poaceae or other herbs at the very early stages of development. This arrested succession can be caused by shrubs that can reach a given low height more easily than trees, thanks to their shrub biomechanic strategy (Chapter 2). Lianas and early successional trees with low economic value may play a part as well. These situations may require random events or human assistance to break through the dense canopy of these faster-growing plants that have a smaller maximum height. Succession can also be halted by fires or browsing, leading to a stable nonforest state (Chapter 3). Unwanted plants can be manually weeded, but this can be expensive, and to be profitable, any investment early in the rotation should lead to much higher revenue at the final felling if the interest rate is high (Figure 9.1).

Weird thinking 10.1 Is planned or market economy better when dealing with trees?

Unfortunately, I am often late despite carefully calculating in advance the minutes that I need for each task. After decades of being late, I finally began understanding the problem. The unknown unknowns that surprise me as I am rushing around asymmetrically take extra time and only rarely free valuable minutes. Similar reasons toppled the Soviet economy, because despite careful planning and even considering the known unknowns, there were unknown unknowns for which the planners could not prepare, which sometimes caused terrible problems. Analogous problems threaten the overly confident forester when planning a chain of forest management operations. Even when planting a tree species that is naturally the main early successional species of the area and boosting its growth with soil scarification, weeding, or even fertilization, the weather, soils, and pest and pathogen occurrences vary, and another species may do better in these particular conditions that are impossible to predict. The forester then makes often a second mistake by stubbornly forcing the stand development back to the planned path even if the other species is valuable as well.

In a market economy, the large number of rival companies increases the likelihood that some are well suited even when the unknown unknowns strike. Actually, companies that do well after unexpected events are the ones that should not have had any chances to succeed according to a rigorous analysis. Similarly, in a forest succession, the dozens of tree species in the tropics and the few in the boreal biome include the species and genotypes that tolerate unexpected flooding, browsing, or unusual soil nutrient statuses. Indeed, a well-tended plantation often accumulates biomass less than a natural succession, but this is not always the case, and comparing plantations and natural successions is a topic of active research.[17]

How to practically apply the laissez-faire policy with trees? First, if the objectives do not include wood production, then often the best strategy is to apply a complete free-market policy and simply let nature take over. This may sound obvious, but it might be less so to a forester educated in a culture where noninterference with natural development is seen as a sign of laziness or a lack of initiative. Indeed, millions of hectares of plantations are unnecessarily managed in much the same way as those intended for wood production, despite having no wood production objectives[18]. Second, even when wood production is important, natural patterns can be actively pursued, such as by planting a mixed forest. Third, when things go unexpectedly, the new management future should be planned based on the current situation. For example, in clear-cut forestry, consider allowing the understorey from natural regeneration to develop into large trees instead of clearing them and planting the same species.

Similar to agriculture, a rapid change in growth conditions can tip the balance so that the wanted plants end up competing with the unwanted ones that have already established on the site. Inundation, like in rice paddies (Weird thinking 9.1), is rarely possible in forestry, but other sudden changes are possible just before sowing or planting timber trees. The site can be prepared mechanically, such as with an excavator, or burned, thus killing or at least slowing down the development of many unwanted plants. In simple even-aged management with a clear-cut at the end of the rotation, the sudden change in light conditions that comes without any extra effort may not allow many weeds that are successful in the open to hold on in the shady forest before the clear-cut.

When the wanted trees reach a metre or two in height, they start shading the ground so that the sprouts of even the fastest shrubs or other tree species considered weeds may not be able to compete any longer if cut down. Similarly, positive feedback may enlarge tiny height differences among wanted tree species, as the shorter trees obtain less light, lose the competition in height growth, and may die out if they are shade tolerant.

Some management regimes do not incorporate any silvicultural interventions from the time the grown trees become dominant plants until the final harvest. However, sometimes fertilizers are applied and thinnings are performed quite frequently. Commercial thinnings have a double objective of both obtaining valuable timber while leaving trees that are able to grow to rapidly increase in value. Generally, trees grow the faster the more they have space around to do so, but obviously growth per unit land area depends on density of trees.

Precommercial thinnings are normally done from below, meaning that mainly the smallest trees are removed, but commercial thinnings are often performed also from above, to obtain more revenue from selling the harvested trees. The question of why thinnings are conducted can be turned around to ask why so many seedlings are planted that thinnings become necessary. The excessively large number of small trees that necessitates precommercial thinning typically results from natural regeneration or direct sowing. However, larger numbers of seedlings are planted in many management regimes than what can grow to the size expected in the final harvest. Greater tree density increases the overall productivity of trunk wood, at least during the early stages, as the leaf area per unit land area is larger. However, if the thinned trees are left to decay, their biomass should not be counted in the productivity that is valuable. Then, additional planted seedlings can be valuable if a large proportion of seedlings die early or to increase the trunk quality. Crowding reduces branch thickness and speeds up self-pruning of the lower branches, both of which will increase the quality of the obtained lumber. The number of commercial thinnings per rotation ranges; it is often one or two, but even three is possible. More frequent thinnings would boost productivity, but harvesting is costly, and the logging machines may damage the remaining trees.

Thinnings from above, or in other words removing the largest trees due to their highest value, can be considered an intermediate form between the more common plantation forestry based on thinnings from below and so-called continuous cover forestry that is not based on rotations, where the final harvesting is followed by the establishment of a new stand. Instead, thinnings in continuous cover forestry trigger the development of natural regeneration, and the age and size distributions of the stand remain similar after each thinning. Regeneration can be continuous, and all ages can therefore be present, but cutting the best can also lead to a deteriorating trend, where only poor-quality trees are left after a few harvests (Weird thinking 10.2). Regeneration of the tree species without an open phase obviously requires some shade tolerance from the species in question, but on the other hand, much or even most of the available light is always used by the trees and therefore theoretically could lead to greater productivity. As there is no sudden drastic change in the light environment, and soil scarification is difficult without harming the roots of the remaining trees, favouring another species over the strongest competitor is challenging, and even then, too few seedlings may establish. On the other hand, poor regeneration is not an economic disaster as it is in even-aged management, where a period is dedicated only for the establishment of seedlings. Thinnings from above may cause severe shock to the remaining trees, as they are suddenly exposed to much more light, which may reduce their light utilization efficiency or cause desiccation due to increased transpiration. In addition, the proportion of the valuable heartwood may be smaller in continuous cover forestry, as the dominant trees have greater leaf area and therefore greater requirements for the water-transporting sapwood.

Weird thinking 10.2 Can only good ones be taken away forever?

Obviously, a monoculture with continuous seedling regeneration and otherwise identical trees of various ages and sizes can be sustained eternally by selecting and cutting the large trees. However, the situation is more complex in multispecies systems. I know how a forester thinks, and I have spent hours upon hours observing how sheep forage in the meadow enclosing our house in Porvoo, Finland. In many ways, we foresters and sheep think similarly and target certain species. The proportions of total biomass of the targeted species decline immediately after logging a tropical rainforest managed for timber or after sheep grazing. However, the subsequent patterns following the initial disturbance differ.

We foresters aim for species with large and straight trunks and decay-resistant wood, which increases the value of the lumber and lowers the risk of a harvested trunk being partially decayed. All these traits are typical for slow tree species. Fast, less valuable species benefit from both mechanisms; they increase immediately because they are not harvested but they also do better in the disturbed logged forest. So if harvesting valuable trees is the only intervention, the sole way to do this sustainably is to harvest only very large trees and wait dozens of years or even centuries between loggings,[19] and even then the original forest structure and species composition are not quite reached.

Sheep aim for herbaceous monocotyledon and dicotyledon species that invest only little in chemical defence. These fast species grow quicker because they invest less in defensive compounds, and concurrently, they invest less in defensive compounds as they grow fast and reproduce young. They benefit from disturbance, potentially even from the sheep eating them. Whether a given herb species favoured by sheep declines or not depends on the strength of the two mechanisms; that is, the initial biomass decline due to grazing and the longer-term floristic impact. Rangeland managers must therefore make more complex decisions than we foresters or sheep. Depending on the exact set of species and conditions, the pasture can improve or deteriorate with increasing grazing. In theory, the right level of grazing could sustain an excellent pasture forever in invariable conditions. However, some fast species may not be favoured by sheep while some slow species may be, and fodder productivity can often be increased with complex management interventions ranging from simple pasture rotation and variation in the grazing pressure to drastic measures resetting the harmful succession, such as burning or tilling the soil.

There are several intermediate forms between traditional plantation forestry with thinnings from below and continuous cover forestry with all ages present. For example, very small clear-cuts can be made to enable the regeneration of light-demanding species, and this may still be called continuous cover forestry. The management regimes based on cohorts represent another intermediate form. Instead of the continuous regeneration in stereotypical continuous cover forestry, regeneration could happen only directly after thinnings, and this could lead to a structure with two or more cohorts. A cohort-based structure is also possible in rotational forestry, such as if a fast-growing light-demanding species and a slow-growing shade-tolerant species regenerate after a final felling. It is important to openly compare the various silvicultural systems and to avoid the common trap when doing so (Sceptic's question 10.2).

Sceptic's question 10.2 How not to compare even-aged management and continuous cover forestry?

One of the largest silvicultural questions in many regions where wood production is the primary objective is whether continuous cover forestry should be practised instead of the more common rotation-based even-aged management with final fellings. I assume here these final fellings to be clear-cuts and the continuous cover forestry based on the selection cut of individual trees. Many correctly argue that much more wood is obtained from a clear-cut. But a selection cut is better for most other benefits. Most people are daunted by the drastic change in the scenery due to clear-cutting, and the erosion and nutrient run-offs are much higher after a clear-cut than a selection cut. Similarly, the carbon storage of a forest collapses in a clear-cut, but the change is much lighter in a selection cut. Such comparisons are common in the popular press and influence

continued

continued

---

the perceptions of forest scientists, at least until they potentially begin researching the topic and think more carefully.

Comparing one clear-cut with one selection cut is terribly misleading, as the frequency of selection cuts in continuous cover forestry is much higher. If selection cuts occur five times more frequently than clear-cuts, we should compare both the positive and negative impacts of one clear-cut relative to five selection cuts. In a landscape in which clear-cuts are used, the area of recent logging is only one fifth of the logged area in a landscape with continuous cover forestry. Probably the revenue from one clear-cut is about five times greater, but so are its environmental impacts. Therefore, the revenue and environmental impacts are both about equivalent per unit time, and both management regimes are roughly equally good, and the focus should be on the much more subtle deviations than based on the misleading comparison of one clear-cut to one selection cut. These deviations result from various mechanisms. First, yield may differ, and both directions are possible. Second, management costs are typically higher in clear-cut forestry, with a need for planting. Third, the logging cost per unit volume of harvested wood is cheaper from clear-cuts without need to worry about damaging the remaining trees. Fourth, forest structure variation is much smaller with continuous cover forestry, and this can impact biodiversity both ways depending on what kind of habitat the focused species require. Fifth, the five-fold change resulting from a clear-cut can be a much more than five times greater problem when considering scenic beauty or value of the forest for recreational purposes.

If clear-cut forestry is based on commercial thinnings in addition to clear-cuts, the assessment becomes somewhat more complicated. A fair comparison could be between five selection cuts relative to two thinnings and one clear-cut.

It is interesting to consider the optimum number of tree species. When the focus is simply on wood production, only one species is normally better because the best species can be chosen, simple management strategies can be developed for this species, and economics of scale help marketing and selling. However, this is not always the case. Above, I mentioned the two-cohort system with two species, which can be clearly better financially, as the shade-tolerant species uses light levels that are useless for the light-demanding species and the latter develops rapidly during the initial phase, and not much light is wasted on nontree plants.[13] Sometimes tree species may have similar differences also in their below-ground functioning related to water or nutrients. One species could have shallow roots, while the other may focus on deeper layers. Increasing the tree species number in forestry also lowers the risks, for many reasons.[14] First, the risk of total failure is lower, as pest and pathogens are often host specific and abiotic disturbances, such as drought, are more detrimental to some species. Second, biotic disturbance agents may be unable to reproduce effectively if their host trees are far apart. Third, damaged or dead trees may spread more evenly within a stand of several tree species, sometimes even mimicking intentional thinnings. Fourth, in climates where tree growth is slow, market prices may vary, and growing several species lowers the risks similarly as investors in markets seldom invest only in one company. With biodiversity objectives, having multiple species becomes a much better strategy because they provide more ecological niches for organisms, and therefore the choice of the number of species often depends on the weight placed to biodiversity in the optimization process.

11

Climate Change Mitigation

A large share of the ecological research and forest science conducted in the past couple of decades has been related to climate change. Most experts seem to agree that forests are important for climate change. However, many convincingly argue that more trees should be harvested to mitigate climate change, while many say exactly the opposite and claim that forests benefit the climate most when they are not touched.[1] Both groups cannot be right.

The main cause of the current global anthropogenic climate change, or just "climate change", is the combustion of fossil fuels. When coal, petroleum, or natural gas sequestered in the previous hundreds of millions of years is burned for electricity, kinetic energy, or heat, the stored carbon flows into the atmosphere and forms carbon dioxide. Carbon dioxide is a greenhouse gas: a gas that absorbs a greater proportion of radiation leaving the Earth than of inward radiation, thereby warming the globe. So the more carbon that moves from the fossil fuel pool into the atmospheric carbon pool and forms carbon dioxide, the stronger the greenhouse effect and the warmer the Earth.

An important portion of climate change has been caused by carbon emissions from ecosystems.[2] Land use, disturbances, and management influence ecosystem carbon density, that is, carbon stocks per unit land area. Human action that has lowered carbon density has increased atmospheric carbon dioxide, similarly to the burning of fossil fuels. However, unlike the burning of fossil fuels, human action can also increase ecosystem carbon[3] (Figure 11.1). This has been the only feasible way with the current technology to mitigate climate change at a large scale: not by reducing emissions, but by reversing them.

Ecosystem carbon can be divided into two subpools: biomass and soil carbon. Biomass refers to the dry mass of living organisms. In forest ecosystems, biomass normally includes both the living and dead tissue of trees, such as heartwood, and may or may not include tree roots or other plants and their roots. Animals are normally not included, but their biomass is minuscule relative to tree biomasses in forests, rendering them unimportant. Nearly half of the biomass is element carbon.[4] For example, a tree with a fresh mass

Trees and Forests of the World. Markku Larjavaara, Oxford University Press. © Markku Larjavaara (2026).
DOI: 10.1093/9780197757109.003.0011

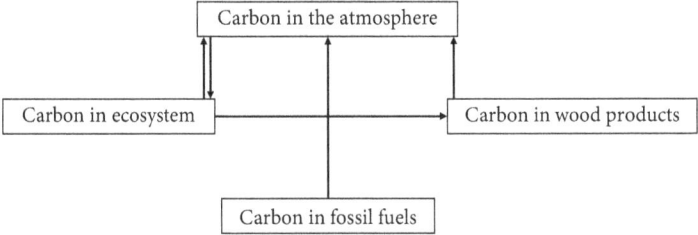

Figure 11.1 Representation of carbon pools and fluxes relevant when discussing the role of forests in climate change mitigation. Ecosystem carbon can be divided into above- and below-ground biomass, dead wood, litter, and soil carbon.

of 1 mg or 1,000 kg may have a dry mass of 500 kg, approximately 250 kg of which is element carbon. The emissions from burning this exemplary tree would equal 917 kg carbon dioxide after multiplying by 3.67, as there is more oxygen than carbon in a molecule of carbon dioxide.

Soil carbon is quite often considered in land-use–related calculations. However, as it changes slowly after land-use change and its quantification is difficult, it may often be better left out. Organic compounds in soils contain variable proportions of carbon, and measuring their mass is not easy, as they are normally mixed with mineral particles. Therefore, a concept analogous to the biomass of living plants is not used with soil carbon, but stocks of element carbon are typically reported instead. Researchers too frequently report soil carbon values without specifying the depth for which the estimate is given. Thicker layers of soil obviously contain more carbon than shallow ones, even though carbon content typically decreases downwards.[5] Overall, in forested ecosystems, density per unit area of soil carbon down to a 1 m depth is normally similar to the density of biomass carbon.[6] However, when trees are stunted or when soils are organic, as in waterlogged peatlands, carbon in soils can easily exceed that of trees even in an old-growth forest. Therefore, globally large areas in the boreal forests with plenty of peatlands have more carbon in the soil down to a 1 m depth, while tropical rainforests have much more carbon in their biomass.[6] Woody debris, mainly from dead trees but also from the branches of living trees, may be a significant component of the ecosystem carbon pool,[7] but litter is less significant,[6] and it varies considerably depending on temperature- and moisture-dependent decomposition and on input seasonality.

To quantitatively understand the importance of past, present, and future land-use decisions concerning climate change in addition to the detailed information of various land uses, we need to have a general framework for the calculations. The frameworks are usually based on fluxes or pools (Figure 11.1). With a flux-based approach, the focus is on carbon emitted or sequestered per unit area and time, or per unit area for a given action. For example, the climate change impacts of deforestation could be 100 mg per hectare of carbon emitted to the atmosphere (Sceptic's question 11.1). Sometimes such emissions are communicated using carbon dioxide as the unit. With a pool-based approach, the calculations are built directly on carbon densities. Both approaches can be used correctly, and the same increase of 100 mg atmospheric carbon could be computed if forest carbon density in a landscape equals 250 mg per hectare and decreases to 150 mg per hectare after deforestation. Much of the confusion on whether trees should be harvested or not to mitigate climate change is caused by wood usage advocates focusing their thinking on the fluxes (i.e. arrows), while forest conservation proponents think about the pools (i.e. boxes); (Figure 11.1). This difference does not directly influence the results but hinders communication between the groups and hampers discussions regarding the assumptions of wood products that cause these opposing views on forest usage, as I will explain below. Personally, I find the flux-based approach more natural when focusing on fossil fuel emissions, but concentrate on the pools when thinking of land use. Carbon is an element, and it cannot be changed into other elements; therefore, it is easy to picture the pools and to understand that if one pool becomes smaller, one or several of the other pools must become larger. Another similar source of confusion traces from people focusing on variable scales. Again, with the same assumptions, the conclusions should be identical at the large scale or when summing up small-scale computations for the same large area. However, in practice, scientists and other experts make different assumptions at the stand and national scales.

Sceptic's question 11.1 How large emission levels from deforestation?

Many forest scientists prefer to begin their funding proposals by stating that forest loss causes a given percentage, such as 12%, of global anthropogenic carbon dioxide emissions.[19] I admit that I have also done this to emphasize the importance of forests. However, such estimates may be misleading, as they depend on many assumptions[20] that are not clear to the reader and sometimes not even to the author. Again, embarrassingly, I confess my guilt.

The deforested site must, by definition, be a forest prior to deforestation but not a forest any longer after deforestation has occurred. The percentage of deforestation therefore depends on the definition of a forest. More dangerously, this percentage also depends on assumptions related to spatial and temporal scale. Most would probably agree that an area where a single tree is harvested and replaced by new seedlings within the year is forestland all the time, and does not contribute to deforestation. Should a 10-hectare stand that is planted with seedlings after half a year be included in deforestation computations? What about a half-hectare stand that degrades gradually over a period of 20 years and remains treeless for 10 years? Similarly, the results are sensitive to the definition of "anthropogenic". These definitions determine how large the emissions from land use caused by us humans are, and how large the mysterious "natural terrestrial sink" is.

The definition of "deforestation" may determine whether a steady-state landscape without net change in biomass or forest area, but with rapid counterbalancing dynamics, is reported as a stable landscape or as one experiencing catastrophic deforestation. Deforestation estimates are normally based on remote sensing methods that are good at detecting rapid decreases in biomass. Increases are naturally slow and harder to detect by satellite or with the naked eye, and therefore receive less attention. Ideally, we should treat changes in both directions equally and mainly focus on net changes. I understand that numbers describing dynamics are interesting, such as the rate of gross deforestation and resulting emissions. Not least, this interest helps gain an idea of their mitigation potential. However, even when the definition of deforestation is clear, these numbers may still cause confusion, as much of the forestation is possible only due to recent deforestation. Deforestation creates the potential for forestation, and presumably the wider the definition of deforestation, the greater the share of deforestation-dependent forestation. Stopping deforestation would not only halt emissions from deforestation but would also weaken the sink related to forestation. Forest lovers dislike deforestation but they should love one of its consequences, as it creates the potential for forestation.

The critical assumptions that are always made but often not discussed are concerned with the fate of harvested wood and the forest's response to logging. Natural resources have traditionally been classified as renewable and nonrenewable. Wood has been classified as a renewable because carbon density nearly always recovers within decades to centuries after harvesting. However, this division into renewables and nonrenewables may be misleading when focusing on the acute climate crisis, as a recovery time of a few decades is long if serious mitigation is attempted within a couple of decades. Instead, it may be better to visualize natural resources along a continuum of renewability, with fossil fuels at one end, with full recovery taking up to hundreds of millions of years, and peat from organic soils recovering in millennia, followed by wood. Any harvesting decreases the carbon in forests but increases it in wood products. The net impact then depends on the magnitude and duration of these changes, and on the optimization horizon that can again be dealt with by using interest rate.[8]

As on several occasions in the preceding chapters, defining whether a given scenario is good or bad is difficult even if the objectives of land management are clearly defined, unless the scenario or several scenarios relative to which the comparison is carried out are defined first (Sceptic's question 11.2). For example, to examine the impacts of wood harvesting, we could compare a scenario with annual harvesting from a forested landscape to a scenario with no harvesting. As explained above, the no-harvest scenario leads to more carbon in the forest and is therefore better for climate change mitigation when the focus is only on the landscape. However, this conclusion may change if carbon in wood products is included. For example, if all carbon is conserved in wood products, such as log houses made from round unsawn logs, their construction mitigates climate change assuming that wood removal accelerates wood accumulation in the landscape. However, it is typically not a good starting point to assume that carbon remains in wood products because only a small portion of a harvested trunk actually ends up as lumber, and even lumber is surprisingly short-lived. In other words, carbon generally remains out of the atmosphere more efficiently in a live tree or as dead wood in the forest rather than in processed timber products.[9]

Sceptic's question 11.2 What should apples be compared to?

It is often said that we should not compare apples and oranges. It is even worse to just say that apples are better, as we do not even know the reference point. Even though this sounds extremely stupid when discussing fruits, labelling an action as good or bad without knowing the reference is done all the time in forest science and environmental management.

If you skip a short drive, you might consume 1 litre less of petrol and avoid emissions of 830 g of carbon into the atmosphere. However, the impact pathway in the complex world never stops, and your reduced consumption lowers market prices, encouraging others to buy more. Yes, this impact is minuscule, but so also is the initial drop in global carbon emissions from the lowered consumption by 1 litre. Additionally, the tiny drop in market rates not only pushes others to consume more, but it also discourages oil exploration and extraction. So it seems that after all, your initial decision lowers carbon emission but less than 830 g. However, you need to also consider what you might do instead of driving; that is, what is the point of comparison. If you ride a bicycle, your body consumes more energy and you need to eat more, and the complexities in the land-use sector are far more challenging than those in fossil fuel use.

"Ecosystem services" is a buzzword confusingly used for both products and services obtained from ecosystems. Scientists have reported estimates on the monetary value of ecosystem services. For some, such as the pollination service for agriculture, the reference without this ecosystem service is easy to envision because an ecosystem without any pollinators is possible to imagine. However, for some services, such as "carbon sequestration", there is no natural reference to be compared to.

Similar challenges hinder carbon and water footprint computations. Many have read about water transpired from natural forests being vital for ecosystems far away thanks to atmospheric rivers and the same water raining down elsewhere. You have probably read about water footprints as well (though likely not in the same paragraph). Somehow a water molecule transpired from a tree in a natural forest is considered good and one from a coffee plant is considered bad. Nobody talks about the terrible water footprint of a forested national park.

Understanding what the point of comparison is and what the borders of the examined systems are is difficult from outside of the specific research community. Therefore, it is important that political agendas do not bias the recommendations.

Because impact chains in the world usually never stop, it is important to consider whether the boundary of the system under examination is at the edge of the examined forest landscape, whether it includes both forests and wood products, or whether a wider system should be considered. Several potential mechanisms may be influential. For example, when new wood products are made, old ones are more likely discarded, reducing carbon out of the atmosphere relative to a modelling exercise without this push effect. Spatial or temporal leakage is another complicating mechanism. When a tree in a given location is not cut, another tree is more likely cut in another location or in that same location at a later date, as initially avoiding the felling increases demand and therefore prices. Substitution is the third potentially important complicating mechanism.[10] Some wood products may substitute fossil fuels or steel as the construction material, and wood product usage may therefore mitigate climate change, even if examining a smaller system indicates the opposite. Because of these complexities, most actions can be considered to both accelerate or mitigate climate change, depending on the assumptions made. Therefore, we should be particularly cautious with calculations made by lobby groups or by researchers with close connections to such groups.

It is interesting to consider how forests and areas containing both forest and nonforest should be managed to optimize climate change mitigation and other objectives. At a small silvicultural scale, there are many similarities between the traditional objective of producing wood and the novel one of mitigating climate change. A poorly growing young stand is undesirable for both objectives. For example, weeding a young plantation with seedlings stagnating under herbs or shrubs is naturally beneficial for wood production (Chapter 10), but also good for climate change mitigation independently of whether the plan is to continue with traditional management but perhaps with a somewhat longer rotation period or to leave the mature trees unharvested to boost ecosystem storage. Similarly, cutting down lianas in a mature tropical forest is beneficial for both wood production and carbon storage.[11] However, optimal management strategies may be very different related to other situations. For example, the optimization management of a mature forest with a significantly higher weight for carbon than for wood productions results in an old-growth stand without harvesting and in a significant increase in carbon stores relative to managed forests.[12] When climate change mitigation is the dominant objective, the old-growth biomass that that species will reach and how rapidly this happens are critical to selecting the appropriate tree species. Energetically (Chapter 7), old-growth biomass for a given species depends on how much biomass the tree species can maintain with a unit of energy, which in turn is expected to depend on wood density, for example, and the levels of secondary compounds protecting the trunk from pathogens. Many nonpioneer plantation species are likely to have large maximal biomasses.

Considering which scenario is better for both climate change mitigation and wood production requires assumptions on how harvesting impacts carbon storage and wood product pools. Generally, boosting wood production without greatly influencing forest carbon storage or boosting carbon storage significantly without harming wood production considerably should be beneficial in a joint optimization. A common approach would be lengthening the rotation period in an even-aged production system or even switching to continuous cover forestry,[13] which works with a shade-tolerating tree species, even with a high average biomass. Alternatively, a management regime with long-lived large trees grown sparsely, and rotations of rapidly growing timber species with at least some shade tolerance underneath could be used. A similar arrangement is common in coffee plantations and has been estimated to a cost-efficient way to boost carbon stores.[8] Very large trees may have a very high ratio of carbon stored as biomass to light absorbed, as the maintenance energy consumption of an individual tree increases less than its biomass.[14] Therefore, huge trunks may be supported by a relatively small crown, which shades much less than the crowns of smaller trees that combined have the same biomass. When similar optimization is performed over a large area with climatic variation, the temperature-caused variation in tree physiology (Chapter 8) leads to interesting patterns (Chapter 14).

Climate change mitigation is an extreme service in the ecosystem services spectrum, as the cost of diverting from otherwise optimal management is paid locally but the benefit is shared by the global population (Chapter 13). As the cost seems initially minuscule relative to the benefits, various mechanisms to motivate and compensate have been discussed and initiated. As the difference between the global and very local scales is so extreme, many practical solutions would be divided into steps involving separate mechanisms at the international and local levels and in between. Mitigating climate change via land management is very cheap for the landowner compared to other discussed mitigation options,[8] but the obstacles in organizing the payments may still make this very difficult or even impossible at a low cost and at a large scale.[15] Incentivizing those in charge of land use may be much more challenging than initially thought. Traditional cultures, often with strong rights, taboos, and expectations of land use, and more modern practices, such as agricultural subsidies incentivizing decision-makers to keep carbon densities low, make significant increases in carbon density difficult. Experts from rich countries may optimistically estimate huge opportunities for their carbon programmes, but these are often caused by naivety concerning the complications that experts foresee in their own cultures yet fail to anticipate in a foreign country.

Developing a fair system for quantifying a compensable amount is one of the major challenges. Paradoxically, even though quantification of the actual carbon density trend with a compensation may be difficult, it is even more difficult to understand the hypothetical business-as-usual or reference scenario, or what would have happened without compensation. Comparing the development relative to the reference scenario seems the best approach, as a slowdown in emissions or speeding up of carbon sequestration to the ecosystem should be compensated equally if their climate impacts are equal. However, at an international scale, many countries have claimed that they should not be forced to perform further efforts if their carbon sequestration trend has been positive even without carbon mitigation objectives. A similar ethical question from everyday life would ask whether a person should be rewarded for doing something good even if they were unaware that what they were doing was a good thing. Another equally puzzling ethical question concerns countries that are at various stages of the forest transition curve (Chapter 9). Should poorer countries that have not yet experienced rapid deforestation be allowed to clear their forests in the future, even if we now understand the significance of its carbon impact? As most countries have already gone through this phase, should countries that have not experienced it be allowed to do so despite current understanding and agreement on the importance of forests in climate change mitigation? This question is similar to an everyday situation where someone first does something bad without knowing it is bad, and then another person wants to do the same after the negative impacts of the action have become evident.

So far, my focus in this chapter has been on the climate impacts of carbon dioxide and associated other changes. However, other greenhouse gases, such as methane and nitrous oxide, are also influenced by land use but are mainly related to agriculture and less with forestry. However, methane emissions from wetlands can be important, and rewetting northern peatlands that were drained to boost tree growth for forestry increases ecosystem carbon and should therefore mitigate climate change, but as the rewetting increases methane emissions, the impact is the opposite in the short term.[16] Nitrous oxide emissions may be important in some intensive plantations with nitrogen fertilization and are worth considering when climate change mitigation is a major objective. The pool-based reasoning that I use for carbon and atmospheric carbon dioxide cannot be used for methane and nitrous oxide, as they are relatively rapidly broken down into other compounds.

In addition to greenhouse gases, forests influence climates in many ways. Water vapour is also a greenhouse gas, but other complex mechanisms related to how trees influence water cycles have a more direct influence on climates. Albedo—reflectance from the Earth's surface—varies depending on land use, and increasing forest area at high latitudes has been suggested to warm the Earth despite carbon dioxide removal. This is caused by the decrease in albedo, especially in the springtime, when solar radiation is already important but is reflected largely back from a snow-covered nonforest area. Taking this into account in climate change mitigation would favour increasing the carbon density of areas that are already forested, especially with evergreen trees, rather than the afforestation of nonforests in northern latitudes. An albedo change caused by forests is much less important in the tropics, but comparisons with greenhouse gas emissions are tricky (Sceptic's question 11.3). Finally, forests also influence climates by emitting and influencing aerosols and other organic compounds in the atmosphere.[17] This impact may be large, but research concerning the topic is in its early stages and we do not even know whether the net impact of these emissions from forests is warming or cooling.[18]

Sceptic's question 11.3 Should we only focus on minimizing radiative forcing?

The impacts of other greenhouse gases are often expressed in carbon dioxide equivalents. Such simplifications may be misleading, as carbon dioxide impacts the biosphere directly and these gases have variable lifespans, and therefore their equivalencies depend on the time horizon that can be expressed with interest rates. For example, if a given emission

continued

continued

of carbon dioxide is as bad as an emission of methane in the upcoming decade, then the carbon dioxide causes more harm in the upcoming century. With very short-sighted optimization and high interest rate, methane is worse than the carbon dioxide emission, but this reverses with lower interest rates.

Comparing greenhouse gases to the other climate change impacts of forests is unfortunately even more challenging. These comparisons are typically done based on radiative forcing, which, when positive, warms the Earth. Alternative scenarios can then be compared, and two options causing equal radiative forcing can be concluded to be equivalent in their climate change mitigation potentials, but such a supposition may be dangerous. Greenhouse gases spread rapidly and evenly around the Earth. However, the climate impact of a changing albedo is completely different, as the main impact is local, and the physical processes differ entirely. Therefore, offsetting the increase in radiative forcing caused by carbon emissions from burning fossil fuels with a small-scale geoengineering project that increases the albedo may have harmful impacts. Even if the increase in global mean temperature could be offset, the spatial, seasonal, and diurnal temperature variations would still change, as would all the other meteorological variables such as precipitation. Even though the strength of climate change is typically expressed as warming of the global mean temperature, most dangers of climate change are not directly related to this warming but to all the other predictable and unpredictable changes. Therefore, a carbon sink needs to be established to perfectly offset the damage from carbon emissions. Hence, it may be confusing to group both projects impacting albedo and carbon under geoengineering and to compare them based on their radiative forcings, as is unfortunately often done.

Are those right who argue that climate change mitigation should be carried out by harvesting more trees to substitute fossil fuels and building materials and by increasing the carbon pool in wood products? Or, on the contrary, are those individuals correct who say that we should focus on forest ecosystem carbon pools and use less wood? The shorter the planning horizon, or in other words the higher the interest rate, the fewer trees should be harvested, and the scenario focusing on forest conservation seems more justifiable with typical planning periods for the future. However, these topics are heavily politicized, and due to computation complexity, it is difficult to pinpoint researchers who are not biased because of proximity to lobby organizations or other scientists with links to them (Weird thinking 11.1).

Weird thinking 11.1 Why does considering climate change mitigation only slightly impact forest management recommendations?

The disagreement of whether to log more or fewer trees to mitigate climate change is not unusual or unwanted in science. On the contrary, science benefits from active debates, as those are encouraged and common especially in Western European and North American countries with less conformist and conventional cultures than in the rest of the world,[21] and which also receive most of the Nobel prices in sciences. Correspondingly, it is common in science that researchers lock their views early on in their career. Physicist Max Planck even argued that advancement in science happens via funerals and not through the conversions of minds.[22] To me this is generally an exaggeration, but not so with forest management–related research. Worse still, even the new objective of mitigating climate change has not changed these views during the 30 years that I have followed forest debates.

In the 1990s, the views concerning Finnish forest management were already polarized among scientists. The block closer to forest-based businesses argued for intensive use, while those closer to environmental NGOs requested more forest conservation. In the completely novel situation with a significant weight placed on climate change mitigation, both blocks remain immovable and use climate change mitigation to declare why the management that they were promoting already in the 1990s should be followed. I also do not believe that dishonesty would be playing a major part, but rather that most people in both blocks write and say what they truly believe. Instead, the polarization is possibly due to the human tendency to absorb more information that supports prior thinking (i.e. confirmation bias), and it is boosted by information cascades resulting from social processes. Similar processes lead children to

continued

continued

believe in Father Christmas and people of all ages to believe in gods. These mechanisms are especially strong when individuals like their prior beliefs.

On the global scale, advocates of intensive logging to mitigate climate change have not had much visibility, but the biodiversity conservation block is similar as its Finnish counterpart in the national scale. They talk about the importance of forests and argue for increasing forest conservation, slowing down deforestation and degradation, and increasing forestation. Other environmentalist scientists, by contrast, claim that forests are unimportant; even letting them reach their maximal biomass would not greatly lower atmospheric carbon dioxide levels,[23] and fossil fuel combustion should therefore be reduced with full force.

12

Other Benefits of Forests

The capability of forests to mitigate climate change (Chapter 11) has been discussed widely for only some decades. By contrast, wood obtained from forests (Chapter 10) has been important as firewood and for constructing tools, shelters, and boats throughout human history. However, food has been even more central in human environmental history than wood has (Chapter 9). It is therefore worthwhile to begin this chapter on other forest benefits by examining how trees and forests contribute to human nutrition.

Imagine a forest and then an open area or nonforest in a given climate. The most fundamental differences between them trace back to the differences in the size and lifespan of the dominant plants. Trees are larger than other plants and therefore have a longer lifespan. The longer lifespan is then linked in both ways, as a cause and a consequence, with greater protection offered against herbivores by the structural tissues that the plant is composed of. Humans can consume a great range of plants. The seed is the most typical plant part that we eat. For example, grains from cereals (i.e. herbaceous plants from family Poaceae) directly provide 45% of human energy intake and just three species, rice, wheat, and maize, are by far the most important ones.[1] Seeds are nutritious to humans, as plants store energy in them that is available for the seedling after germination. Many plants develop nutritious tissue around the seed to attract animals to disperse them. For example, bananas and plantains have huge importance locally as stable sources of nutrition and also as important crops in international trade.[2] However, fruits globally represent well below 10% of human energy intake.[1] Seeds contain energy for future seedlings, but when plants store energy for their own future use without reproduction, they often do so underground. Tubers, such as the potato, are important in many regions. Most leaves and stems are inedible for humans, yet thousands of species, such as cabbage and lettuce, are consumed but often only as secondary ingredients in meals; they are important for micronutrient intake but not for energy supply. However, many edible plants do not fit well in the above-mentioned classes of seeds, fruits, tubers, or leaves, and many plants produce various edible parts. Humans have been

Trees and Forests of the World. Markku Larjavaara, Oxford University Press. © Markku Larjavaara (2026).
DOI: 10.1093/9780197757109.003.0012

able to breed many species or closely related species rapidly into varieties that produce a surprisingly wide array of products. An extreme example is the genus *Brassica*, which yields edible leaves, such as many kinds of cabbages, consumable tubers like turnips, edible flowers like cauliflowers, and canola oil and even mustard obtained from the seeds.

Bananas and plantains are harvested from a large herb, and palms producing palm oil and coconuts are potentially considered to be trees (Chapter 2). Angiosperms (eudicots only as defined in Chapter 1) produce numerous fruits; for example, apples, pears, peaches, and apricots in the family Rosaceae are important in temperate biomes. The tropics have important fruit trees in dozens of angiosperm genera and have contributed significantly to human diets in the pre-Columbian Amazon basin (Weird thinking 9.1). The seeds of gymnosperms are normally wind dispersed and therefore do not develop a flesh to attract animal dispersers, but these seeds are sometimes eaten as nuts, such as the seeds of many *Pinus* species. These examples demonstrate how large plants provide human food, but only a tiny part of their biomasses is eaten, normally only the seeds or fruits, and nearly all the most import food crops are small herbaceous plants. Even within the family Poaceae, the herbaceous species dominate food production and woody bamboos have only a marginal importance in diets. Within Poaceae, the important crops are small and have an annual life cycle. The existence of annuals is dependent on abundant seed production at the end of their life cycle. Their yield has been further boosted by breeding and by providing close optimal growth conditions with abundant nutrients and without competition from weeds. Seeds or fruits of small plants are difficult to harvest without killing the plant, but annuals die anyway at the end of the growing season. Therefore, avoiding damaging the plant is not necessary.

Trees are large and have a large biomass to maintain (Chapter 6). Theoretically, with a given leaf area, a system with smaller plants that do not waste energy lifting their leaves upwards produces more if the small plants are growing densely enough so only little light penetrates to the ground. However, the situation is different below-ground. The size of above- and below-ground structures correlates, and larger tree roots are able to obtain nutrients and water from deep soil layers that are inaccessible to small herbs.[3] The ability of trees to pull up nutrients and make them available to small plants is used in agroforestry, in which trees and other plants alternate spatially (Sceptic's question 13.1), and in swidden agriculture, in which trees are first left to grow and are then slashed and burned, typically after a couple of decades. An additional benefit of trees relates to microclimate modifications that in many cases improve the situation and the associated value by reducing wind erosion. Overall, however, managed nonforest ecosystems have been more efficient at providing food directly consumed by humans compared to forests.

The indirect way in which we obtain food from plants involves domestic herbivores, such as cattle and sheep, which first eat the plants, after which humans consume their meat. A much smaller proportion of the energy entering the ecosystem ends up in the human diet, but on the other hand, livestock can potentially use areas that are difficult to cultivate, and in certain traditional production systems the main value of domestic animals has come from their dung used to fertilize croplands. In these systems, livestock can be seen as vehicles transporting nutrients from marginal lands to cultivated fields. Pigs and chicken are feeding or are fed food that is largely also edible to humans, but cattle and sheep mainly eat plant leaves. They can eat them directly from the plants, or the leaves may be harvested elsewhere and brought to the animals. In both cases, theoretically optimal plants have nutritious leaves with little defensive compounds, are small and otherwise easily eaten or harvested, and regrow quickly after grazing, browsing, or human harvesting. All three attributes are related. As discussed earlier (Chapter 3), small plants with short-lived stems have less evolutionary incentive to invest in defence and more to invest in being able to recover rapidly after disturbance. Again, as with directly human-consumed plants, small species in Poaceae are typically the best ones matching the requirements, but they are not necessarily annuals like in grain production. The domestication of climbing or flying animals could have opened the possibility of using large tree leaves as fodder without cutting down the trees, but obviously such animals would have escaped easily in the early phases of domestication.

Instead of domesticating trees with edible parts or animals that could eat trees, an option is to hunt or gather meat, tubers, fruits, nuts, leaves, or honey from a forested landscape. However, pre-agricultural hunter-gatherer population densities have been very low; they rarely surpassed one inhabitant per km,[2] and more typically were only a tenth of that.[4] The highest densities were in temperate or subtropical forested biomes, with much lower densities in African and Australian savannah biomes,[4] perhaps surprising to proponents of the savannah hypothesis (Sceptic's question 9.1). Already prior to industrialization and modern medicine, these numbers were very low compared to intensively cultivated regions. For example, arable areas of Japan had nearly 400 inhabitants per km[2], and other paddy rice-based economies also had high densities.[5] These numbers illustrate the low density of food that was obtainable by hunting and gathering compared to the most intensive agriculture.

In modern times, hunting and gathering for subsistence remains important in certain regions,[6] but they generally fall under the category of hobbies or recreational forest use. For much of the world's forests, the focus of humans is on wood production (Chapter 10), to the extent that all products obtained from forests are often divided into wood or timber and nontimber forest products, including edible meats and parts of plants, such as medicines, fibres, resins, and cork. The importance of services obtained from forests has increased relative to the products. I already discussed the most important service, climate change mitigation potential (Chapter 11), and for the rest of this chapter I will focus on biodiversity, hydrology, and recreation, and lastly discuss managing forests for multiple benefits.

"Biodiversity" is complicated to define, or more correctly there is a huge range of possible definitions. Therefore, it is worth considering why the term is used (Sceptic's question 12.1). First, in empirical studies not all organisms starting from microbes can be easily studied, and therefore the focus is often just on birds, butterflies, or trees, for example. Second, the focus can be on within-species genetic diversity, species richness, or the richness of some other larger taxa. Third, even though the focus is normally just on spatial diversity, the temporal component can also be included. Fourth, the size of the studied or managed area varies. Fifth, the size of the area for which biodiversity is maximized also varies. Sixth, the common division into alpha, beta, and gamma diversity[7] corresponds partly to the fourth dimension above, which considers variable size, but beta diversity is different because it is concerned with spatial diversity but between communities of species. In practice, biodiversity discussions are typically centred on maximizing species richness at the national level or if the area studied has plenty of endemic species, then species richness can be maximized at the global scale, but the taxonomic group (the first dimension) and the area that is managed (fourth dimension) vary widely. These decisions cause huge differences. For example, if maximizing species richness in Sweden is the objective, then southern forests with *Quercus robur* and other southern species are the forest types to protect,[8] but if species richness in the European Union is the focus, then the forest types in Sweden but which are common in Central Europe have little value, and instead boreal forest types should be the focus in Sweden. Another example, related to invasive species, causes similar confusion. A novel invasive species normally increases species richness in its newly established area, at least initially, but invasives risk causing global extinctions and are therefore bad for global species richness.

Sceptic's question 12.1 Why are we talking so much about biodiversity?

The word "biodiversity" can be defined to mean biological diversity. However, when delving into more specific definitions, the options are so numerous that most land management decisions are beneficial to biodiversity with one definition, unless all life forms are harmed. Therefore, it seems astonishing that scientists largely agree what to do in each area to increase biodiversity even without discussing definitions. This is even more astonishing when compared to the forest climate change mitigation discussion, which is highly polarized (Weird thinking 11.1) despite the clear objective of slowing down planetary warming (Chapter 11). The absence of a similar polarization in the biodiversity discussion among scientists can be understood based on social processes. Those participating in the scientific forest climate change mitigation discussion have different backgrounds and therefore form two opposing blocks, while biodiversity researchers form a more unified group with a common cause of promoting the protection of living organisms.

   Biodiversity is a very common concept but is not easy to understand. Terms developed by and used among scientists but also intended for policymakers and the public can arise and spread due to various reasons. Sometimes they can be very helpful by condensing a complex concept into one or two words and thus saving time (Weird thinking 8.1). However, this is not the rationale for using the term "biodiversity," as its great range of possible specific definitions necessitates additional definitions to be used precisely. Like many buzzwords, such as "ecosystem services" and "sustainability," "biodiversity" also has a positive connotation. Plenty of donors are surely eager to fund a project that claims to increases biodiversity and other ecosystem services in a carbon-smart and sustainable manner. Unlike the other three buzzwords, "biodiversity" has several precise definitions, has been used widely already since the 1980s,[21]

continued

continued

---

and is central in the legislation of many countries and in numerous international agreements such as the Convention on Biological Diversity.

Has biodiversity conservation, in a way in which it is normally understood, benefitted from how the word "biodiversity" is used? Probably greatly, as the term is so difficult to understand that experts now have a monopoly regarding what practical measures should be carried out and where, after having first convinced the public of the importance of biodiversity conservation through heavily selective reasoning. If the discussion was on bird species richness in India, the public could better engage and compare costs and benefits of the conservation actions. Currently, with the abstract concept, conservationists have been able to pull decision-making out of the democratic processes, and this has probably been central in the sudden and sustained use of biodiversity as a term.

I have used and will use the term "biodiversity" in this book similarly as most others, focusing on the species number of well-known taxa but excluding invasives.

In many situations, the precise definition of biodiversity is not critical. Theoretically, the biodiversity of a group of species should be higher in ecosystems that resemble the ecosystems in which the species have evolved. For example, an urban environment that none of the potential species has experienced in its evolutionary history is suitable for some but not many of the species. The abundance of large organisms is another feature boosting biodiversity. Thanks to their physical size, large organisms already create more height variation[9] with plenty of variable physical conditions. Trees are such organisms, and even more variation in microhabitats is created if the trees are of variable sizes, species, and health statuses. Therefore, natural forests are generally perfect for biodiversity. In our global study,[10] these theoretical patterns were supported when national experts were interviewed regarding the taxa that they are experts of. Generally, increasing the occurrence of trees and their size (i.e. their above-ground biomass) boosted the value of the land-use class for biodiversity. When this was not the case, the comparison was between a structurally simple plantation and a more complex, natural shrubland.[10] The naturalness of an ecosystem tends to increase the number of species it supports. However, when focusing on endangered species, the rarity of certain natural features or habitats, such as dead wood, also increases their value for biodiversity conservation. These features may have been common in the past but are now less so.

Because trees shade the soil surface in a forest, it is typically moister than in a nonforest. However, as trees are taller than the smaller plants found in nonforests, with a given leaf area per unit land area forests transpire more, assuming other aspects are equal. The greater transpiration of forests than nonforests is further strengthened by trees having deeper roots,[3] which allows for better access to soil water even during droughts. Actually, forests have a slightly higher leaf area relative to land area than nonforests do,[11] further increasing transpiration relative to nonforests. Because forests transpire more, their soils are typically much dryer and less water runs off to downward rivers.[12] This can be good but is normally considered bad. However, other aspects should be considered as well. First, the water transpired from the ecosystem is not lost forever. It will rain down elsewhere, and the more water is transpired the greater the precipitation.[13] Much of this precipitation occurs over oceans and is therefore useless to terrestrial managed and natural ecosystems, yet much of it falls on land. Such mechanisms have been discussed widely related to Amazonian deforestation.[14] Some model runs predict a vicious cycle with less precipitation because of human-caused deforestation, which will trigger drought-caused deforestation and even less transpiration that reduces the precipitation even further. The second complicating factor is related to temporal variation. Along most rivers, even in arid regions, people would prefer less flow after exceptionally wet periods and more flow during dry spells—in other words, a more even flow. Even though forests transpire more than nonforests during the dry season, trees influence soil surface and the soil itself so that infiltration increases. Infiltration during wet days reduces runoff, and its slow release back to the surface in springs at lower locations evens the downstream runoff.[12] The third complicating factor is related to water quality. Very little erosion occurs in stable natural forests, even though much of their area lies on slopes as steep as stability allows, because most of the easily erodible material has already been removed in the past and the runoff water is clear. However, croplands are typically young, and the frequent soil tillage makes the actual top soil structure even younger.[15] Controlling water erosion itself can also be considered an advantage of having a forest rather than a nonforest on a site. However, it is not impossible to consider the positive impacts of erosion, such as the lowered landslide risk and the possibility for downriver farmers to use the often-fertile eroded soil on their croplands in fertile river deltas.

Forest ecosystems are important culturally, often even without the need to visit them.[16] Recreational values have increased with increasing level of wealth. The action of harvesting, such as hunting or berry picking, can be more valuable for many of the above-described products, such as meat and fruits, than the products themselves, especially in wealthy countries.[17] Instead, if the labour cost of harvesting forest products is also considered, the value of forest products is often negative. The stereotype of a beautiful ecosystem in which people enjoy spending their free time varies from country to country, but there seems to be a general tendency of appreciating strongly modified, intensively managed production systems in countries with less time since poverty and famine. By contrast, people in rich countries tend to value natural ecosystems,[18] and on the other hand, systems managed with traditional, perhaps centuries-old practices.[19] Similar differences are visible within countries that I have visited, to where urban and young people favour more natural ecosystems compared to older rural folk.

In practice, managers have nearly always considered several benefits from forests concurrently. In a larger share of the world's forests, wood production objectives dominate, and other benefits are attempted to increase without greatly decreasing production. If climate change mitigation is done by boosting wood production, the objectives align without any conflicts, while carbon storage can be increased with longer rotation periods (Chapter 10). As the future is uncertain and carbon storage could be the main target far in the future, it might be wise to plant and allow natural establishment of slow species (Chapter 3) that potentially develop large biomasses. Biodiversity objectives can be advanced without harming wood production much by using multiple tree species. Increasing dead wood, which has been abundant in the evolutionary history of forest-dwelling species and is therefore critical for many,[20] is problematic because silvicultural systems optimized for wood production attempt to harvest most wood before it begins to decay. Closed-canopy forests are hydrologically similar to each other, but clear-cuts dramatically reduce transpiration. Clear-cuts or less severe forest operations may trigger erosion and lower downstream water quality.

When trying to simultaneously manage more than two benefits, things naturally become more complicated. Optimization is further hindered by experts typically overestimating the importance of their own area (Sceptic's question 12.2). It seems that deforesting in one area and foresting in another is rarely praised locally. Recreational uses often prefer continuity. The creation of a new recreational area only partly offsets the damage caused by the disappearance of another area. For decades or centuries, the new forests have much less value for carbon storage, and it takes even longer for biodiversity value to increase.

Sceptic's question 12.2 Are conservation biologists average citizens who just know a lot about conservation?

I have attended many scientific conferences where a large share of all the presentations have focused on conservation biology, but only once have I participated in a conference focusing solely on conservation biology. There, when talking about an area under threat of increased human use, one of the keynote speakers proudly said, "I did what I could as a scientist, which is I wrote a paper outlining the scientific value of this area". Over a lunch table conversation in the same conference, a scientist encouraged an NGO activist to tell her how she and her colleagues could help in efforts to conserve forests. Is it okay that scientists actively ask lobbyists to influence them?

Conservation biologists argue that their discipline is applied science aiming to conserve nature. This could be interpreted so that the scientists in the conference were thinking that their role is to use all means possible to increase nature conservation. From the funding perspective, their objective would then be to pursue more funding for conservation instead of optimizing action with the given resources. Then conservation biologists studying a forest area would select their research methods to get the wanted results biasing the research. Those involved in practical land management appear to know this. They would be surprised if a conservation biologist would suggest decommissioning a protected area after studying its fauna and flora for years. People expect conservation biologists to argue based on their research for increasing conservation, and other people must then guess the importance of the conservation based on the strength of the argument.

In my opinion this is not okay, and the damage is not done just to funding allocation but more generally the objectivity of science may be harmed (Chapter 16). However, it is difficult to vision how the culture

continued

continued

could be changed towards more objectivity. It is easier to understand how we have come to this. Those who want to study conservation biology are already nature lovers. Their mindset is influenced towards greener prospects by the usual confirmation biases and information cascades (Weird thinking 11.1), in which even small selectivity may lead to severe biases when the message is passed onwards several times.

PART IV
FUTURE FORESTS

13

How Much Forest Is Needed?

During a university course that I taught several times, I asked the students to list ways in which forests influence human economy and well-being. The students have been very creative and have listed dozens of mechanisms, all of them positive except for a few jokes. Obtaining a partial, positively biased set of answers was my intention, as I wanted to demonstrate how easy it is to think about positive aspects when we contemplate forests. My question regarding the ways in which forests influence humans has not been exact. Discussions in all areas of society often lack a clear reference to which the comparison is performed (Sceptic's question 11.2). In my case, it was not clear whether a forest should have been compared to an empty, useless space or to some more realistic land-use alternative for the same location. Instead of a forest, the same area could, of course, be devoid of plants and be occupied by urban or perhaps road development. However, based on land-use changes during recent decades, croplands with agricultural plants and pasture or rangeland with grazing domestic animals are more common alternatives for forests (Chapter 9). Are forests better than croplands and pastures, and should all croplands and pastures be forested?

The students attending my course over the years have not been alone with their positive views of trees and forests. People enjoy reading about trees positively affecting their seedlings (Weird thinking 6.3), and everyone seems to view forestation positively. Wangari Maathai was awarded the Nobel peace prize in 2004, largely for her work in advancing tree planting in Kenya. Nadine Unger, on the other hand, questioned the value of forestation in mitigating climate change, but an esteemed group of researchers rapidly told her she was wrong. According to Unger, some of her colleagues stopped talking to her, and she even received death threats.[1] All of these events point us towards the thinking that planting trees is really good and that all the world's croplands and pastures should be forested.

My grandmother's father was a physician in a town in southeastern Finland in the 1920s. He invested in land, including an abandoned farm, and planted timber species with my grandmother and the rest of the family.

Trees and Forests of the World. Markku Larjavaara, Oxford University Press. © Markku Larjavaara (2026).
DOI: 10.1093/9780197757109.003.0013

Plenty of survivors of the great famine of the 1860s[2] who focused on food production were still active, and foresting agricultural land was illegal. My great-grandfather was fined, but the *Picea abies* seedlings were allowed to grow, and in recent years, I have used my clearing saw on these same plantations to help the second generation of *Picea* seedlings survive among faster-growing angiosperms. This serves as an excellent example of how the current values of the urban, educated, and Western people have not always been dominant, and we should therefore consider the pros and cons of forestation. After all, perhaps not all croplands and pastures should be forested, as we need to eat, and forests are not efficient at producing food (Chapter 9).

We need both forests and cropland, but how much of each? Is the current forest area in each country close to optimal, and if not, why not? Could it be a question of poor comprehension? Uneducated farmers clearing forests who do not understand their own best interests? Or are there conflicts of interest?

To begin a more analytical examination of how much forest is needed, I start with a simple case. Consider a homogenous rural area without any trade across its borders. Forests provide a range of products and services, and nonforest areas provide another set of products. If the whole area was forested only a small amount of food could be produced, and most people would starve to death if population density was high. On the other hand, many potential foods and timber would not be available without forests and trees, and depending on local climate and soils, wind erosion causing soil degradation could be substantial. A small increase in tree numbers would initially be very valuable; the marginal value of additional forest area would be high. Rare fruits and timber would be valued highly, and wind erosion would decrease significantly even with a very low, such as 1%, proportion of forest if the trees are positioned in small patches or narrow bands. In modern wealthy economies, in which only few people have physically demanding occupations and many people enjoy leisurely forest walks, a similar proportion would already allow significant recreational use of transport is easy. However, with a subsequent increase in forest area, the marginal value of forests would decrease and at some point their value would dip below that of the alternative land use, that is, nonforests, representing the optimal proportion of forests. The exact number between 0 and 100% would then depend on the values of the products and services obtained from the forests and nonforests.

If we still envision an enclosed area but with more variability within its borders, we can more realistically picture many of the processes acting in real landscapes. When population density varies, the production of horticultural products that are difficult to store and transport may be optimal close to population centres. The next zone, away from concentrations of people, could be dominated by croplands, with a focus on grains that are easy to store and transport but that still require relatively intensive management. Further away from the population centre, timber plantations and pastures could become dominant, and lastly, natural forests would be located on the fringes of the envisioned area, with sparsest human population density and longest travelling time to urban areas. Such spatial patterns are often found on a national scale in smaller countries or on a provincial scale in larger ones. However, topography and soils normally vary significantly at these scales, and the main cause of the observed patterns may not be transportation distances. Because high human population densities occur in areas with fertile soils due to historical reasons (Chapter 9), the most-productive agriculture could be located close to population centres, not because of the short food transportation distance but because both people and agriculture are where the fertile soils are. Similarly, agriculture is already challenging in the rough topography that is often found close to the frontiers of zones inhabited by groups of people, where the borders of a province or country were therefore formed. These remote areas could be dominated by forests, not because of the long distance to population centres, but again, because of the topography and soils. Plains, with croplands, typically have a finer soil texture that stores more water in the surface layers accessible to small plants. Sandy upland soils, however, can be challenging to small plants with shallow root systems, yet they concurrently promote water infiltration that may lead to reserves deep within the soil that are nonetheless accessible to large trees with deep roots.

The average productivity of workers increases with economic development. The segments of economy in which this productivity increase is rapid will pull in labour from other segments and become more important. These developments are not spatially even, and some countries or regions within countries will develop faster. In labour-intensive sectors of agriculture, such as the production of vegetables, fruits, and berries, increasing productivity significantly has been challenging because many steps are still carried out manually, while tractors and other machinery can be used in grain production, even for harvesting. Forestry has never been very labour intensive, as most management regimes have periods of several years without maintenance, yet in a way it has depended on the availability or price of land. When envisioning an area larger than in the previous two paragraphs and governed by two or more administrative zones, such as countries with various levels of economic development, the less developed one with cheaper labour costs will likely attract the labour-intensive production assuming that transportation to the markets of the wealthier country or other region is not too expensive. Of course, agricultural workers can alternatively migrate to the wealthier country for the labour-intensive season, as Mexican employees do in the United States. If such migrations are not possible and the products are moved across the border, the decline in agricultural area could vacate land for forest in the wealthier country. This pattern can further be reinforced by wealthier citizens with increasingly physically undemanding occupations appreciating the recreational values of forests more, which could bring forests even to areas close to population centres.

The complexity of the geographic systems has increased steadily in the discussion above, yet even the last and most complex situation is far from reality. Comprehending the spatial and temporal difference related to paying the costs and receiving the benefits of land-use decisions is critical for our understanding.

Practices, customs, and legislation vary around the world, but food produced from a given area of land typically benefits people living on or very close to that same piece of land. Agricultural crops are typically produced on private land, and the official or customary landowner is the benefiter. Other food, such as food collected or hunted from the wild or domestic animals in pastures, typically benefits the local people even if they are not the official landowners, which is still often the case in the poorer countries of Africa and Asia. However, even if nontimber forest products can be legally collected or their illegal collection is tolerated, wood is often considered more valuable and is mainly harvested by landowners, who frequently are not local private persons but companies or state agencies.

Services from ecosystems are spatially spread more widely. The hydrological benefits of forests are often enjoyed hundreds or even thousands of kilometres downstream. Similarly, the reduced wind erosion thanks to forests may benefit people at similar distances downwind by decreasing dust levels in the air. The reduction in wind erosion also benefits farmers by boosting soil fertility. The recreational value of ecosystems is typically enjoyed by people within a few kilometres, but it is not uncommon for people in wealthy countries to travel hundreds of kilometres for a hike. In a way, the recreational values of some unique ecosystems benefit people around the world who watch nature documentaries filmed in these ecosystems. The spatial spread of biodiversity conservation value is similar, but more emphasis is placed on the global aspect. People appreciate local biodiversity more than that far away, but they may still value knowing about the great diversity existing in other parts of the world, even without seeing the filmed documentaries. If produced food represents the other extreme, as its benefits are local, ecosystem carbon represents the other. Changes in carbon storage influence global climate change, and for a local climate, a tonne of carbon sequestered locally or on the other side of the world is exactly the same, as carbon dioxide spreads around the globe in a short time. These differences in the spread of how people around the world benefit from products and services from a given ecosystem are extremely important because a decision-maker determining land use and its management does not normally consider the benefits to other people (Figure 13.1).

Analogously, as in space, there are differences in when products and services are obtained for a forest-related investment. A very similar graph as the one depicting the number of people that depend on a spatial dimension (Figure 13.1) could be drawn for temporal dynamics. Food from a cropland can be produced within months from sowing, but timber trees in plantations typically require decades to mature. Grains obtained from annuals and wood from timber species in plantations are harvested at the end of the rotation or cycle. However, harvesting is more continuous in some other production systems. For example, forests may be managed with the continuous cover system, with mature trees harvested once a decade or so (Chapter 10). An orchard with fruit trees takes some years before maturing but may produce for decades, with harvesting occurring annually or even more often. The services obtained from ecosystems are similar to these orchards, but the "harvesting" is truly continuous. For example, an ecosystem restored to a natural condition may take some years or decades to start producing the biodiversity conservation service, similarly as the orchard takes time to mature. Correspondingly, sequestered carbon, which is away from the atmosphere, is a service that takes time to mature when ecosystem carbon accumulates. However, note that the service related to ecosystem carbon and associated payments is normally perhaps illogically envisioned to happen when carbon is sequestered from the atmosphere into the ecosystem and not during the time it is stored away from the atmosphere, which is the actual mechanism

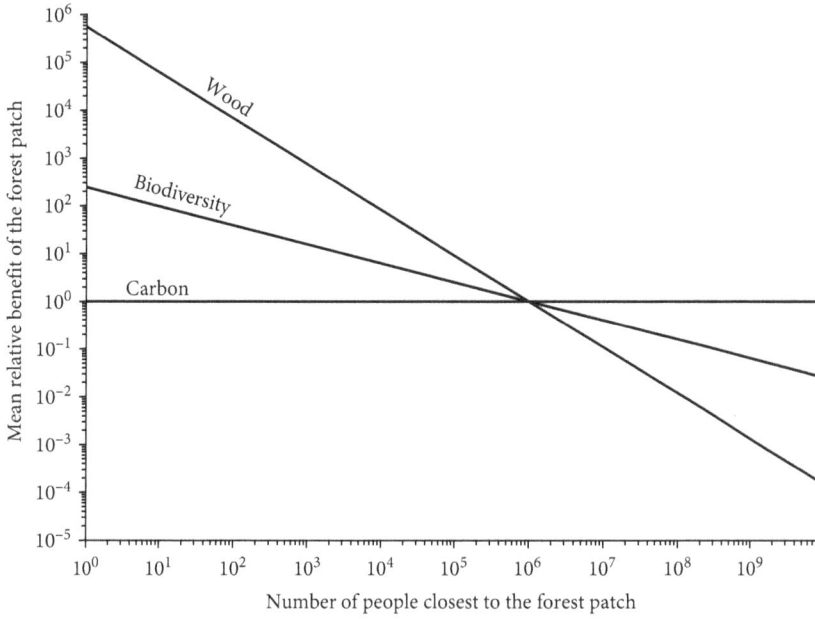

Figure 13.1 Hypothetical example of how the benefits of wood production, biodiversity conservation, and carbon storage to mitigate climate change from a patch of forest could be distributed to the world's eight billion people. I have scaled the three lines, so that the average benefit is one for the one million people closest to the forest patch, corresponding approximately to the national or state scale. If decisions are made at this spatial scale for the benefit of all people within the area, then boosting wood, biodiversity, or carbon are all equally beneficial. However, the right end of the carbon line, indicating the mean benefit for all eight billion people of the world, is approximately 40 times higher relative to the biodiversity line and 7,000 times higher relative to wood production. Note that both axes are on a logarithmic scale.

mitigating climate change (Chapter 11). Analogously, payments related to a fruit orchard would be made when the fruit production capacity increases and all the production would be given to the payer, instead of being made when production happens.

As persons and other decision-making parties have varying time horizons, the optimal solutions will differ even with identical valuing of products and services obtained from ecosystems. I used interest rates to explain what I called the investment strategies of trees (Weird thinking 3.2). In this chapter, interest rates can be used in a more ordinary manner to describe the planning horizon of those making decisions. A short-sighted entity, such as a state or private individual desperately and urgently needing income, should base its decision on a high interest rate. Likewise, if risks are high, for example due to unsecure land tenure, interest rates are high. Then, in modern times with efficient land-clearing machinery, deforesting a landscape, selling the timber, and planting a cash crop with the first harvest within months from deforesting may be an attractive choice. On the other hand, an investment such as foresting an abandoned pasture is profitable only for far-sighted parties with a low planning interest rate.

The temporal dynamics of managing landscapes are further complicated by the evident asymmetry in the speed of changes (Weird thinking 13.1). A mature forest can be converted into a productive pasture in weeks or months, but converting a pasture into a mature forest may take a century and can be very difficult if the ecosystem has switched to another steady state with frequent disturbances (Chapter 3). This can lead to development that is far from maximizing optimality over a prolonged period if the optimal solutions at given points of time are followed. Then, a short period during which the optimization interest rate is very high, such as because of a war or another crisis, may lead to forest destruction, which in normal conditions is considered very harmful. Finally, it is worth noting that even if reforestation is possible after forest destruction during a crisis, the new forest is likely to differ from the original one. If the decision-makers highly value products, such as wood, the new forest is likely to be a plantation that is very different from the original natural forest. The picture is further complicated by intermediate land-use types (Sceptic's question 13.1).

Weird thinking 13.1 Is it okay if net forest area does not decrease?

In the main text of this chapter I explained about the existing asymmetry, where ruining an old forest is much faster than creating a new one and forest age matters when considering the benefits we obtain from forests. In temperate and boreal climates, it normally takes decades before commercially valuable timber can be harvested from a new forest, but growth is faster in the tropics (Chapter 14). Similarly, carbon storage develops slowly over decades. Generalist forest animal and plant species may do

continued

continued

well already in young forests, but usually it is the old-growth species that are endangered,[3] and the very slow development of microhabitats, such as dead wood or the cavities in living trees, may be critical for them. Hydrological and erosion control benefits of forests develop much faster and may not change much after canopy closure. However, a young forest may overall be closer to an open ecosystem than to an old forest when considering the benefits typically attributed to forests, but without the benefits of the open ecosystem.

So, deforesting one area and foresting an equal area elsewhere is a bad deal for decades while the new forests are still young. However, there is one other aspect that I tried to describe in my MSc thesis and in an associated short paper in Finnish,[4] but have not talked of much since then. If you enjoy walking in forests as I do, and normally walk in a landscape of even-aged managements with rotations normally ending in clear-cuts, think of a clear-cut appearing in your favourite old stand and a clear-cut elsewhere magically turning instantly into an old stand. Even if the age distribution in the forest landscape does not change, you probably do not like the changes that occurred within the landscape. People seem to enjoy stability in managed nature, and this is another completely different reason to value discontinuing deforestation more than boosting forestation. Conservationists are conservatives in the traditional nonpolitical sense of the words.

Back from your local forests to global forests. Currently, there are several global forestation initiatives[5] with a lot of fanfare and optimistic declarations of the huge open areas that will be turned into forests. As decelerating deforestation is much more valuable than increasing forestation, we can ask why so much emphasis is placed on new forests. A cynical mind may begin wondering whether the differing time horizons could be influencing the emphasis on increasing forestation instead of reducing deforestation. Projects halting deforestation should show results within a year or two, but forestation takes time and a decade or two after all the fanfare, when the results are turning out to be disappointing, the politicians and others involved have already moved on and are not blamed for being unsuccessful.

Sceptic's question 13.1 Should we favour agroforestry and other hybrid strategies?

Agroforestry is often defined as a management system in which agricultural herbaceous plants are grown among trees that according to some narrow definitions, must provide timber. Some broader definitions include pastures and shrubs instead of trees. If you have read scientific articles on agroforestry, I bet you were at least initially thrilled. The listed benefits range from reduced poverty and improved human nutrition to carbon sequestration and reducing deforestation. A meta-analysis quantitatively summarizing all suitable articles on agroforestry would show a very rosy and very biased picture. Most articles compare agroforests to open monoculture croplands, and many authors are concerned about how farmers do not understand their own best because they do not accept trees on their croplands.

The agroforestry literature suffers from three problems. First, it often mixes benefits from having increased plant species diversity with increased woodiness of the plant species, which have very different impacts. Second, the focus is on comparing agroforestry to croplands instead of using a better reference, that is, a mosaic of forest and cropland patches with a similar mean woody biomass per unit area. Third, understanding that the common optimal solution always differs from an optimal solution for an individual (Figure 13.1) is not discussed as much as it should be.

Forest landscape restoration could be a solution to optimizing land use including agroforestry. However, definitions of forest landscape restoration are so vague that in the preface I listed it among the concepts that I would avoid in this book. In practice, the concept is used for good land use that is planned on a landscape scale and that includes some forest restoration that can be either simply tree planting or more sophisticated attempts to restore a natural ecosystem. I am not optimistic that I would receive project funding with my more precise definition, despite other proposals promising forest landscape restoration having received at least hundreds of millions of US dollars.

To answer the question posed in the title: sometimes we should indeed favour agroforestry. For example, with a shade-tolerant plant like coffee,

continued

continued

---

mixing the crop species and larger trees could be an excellent solution, especially when plenty of weight is placed on carbon or biodiversity benefits. However, much of the existing research on agroforestry offers little guidance in knowing when it is the best strategy. Scientists should use more analytical approaches that distinguish the impacts of having trees in the landscape, their clustering, and those who benefit from them.

The situation is chaotic for anyone attempting to optimize land use and management. Biodiversity and carbon are ecosystem services that benefit people all around the world. Even if these global mechanisms are excluded, it remains extremely difficult to find anything near the optimal solution maximizing the well-being of citizens of a given country. Even if this were possible, the second important and difficult step of directing people to take the appropriate action remains. A simple solution for this could be a planned or command economy directed centrally for the management of land use and resources. In such an administrative system, the production of services and products could be balanced optimally based on best-available information. However, many historical cases have shown this to work poorly (Weird thinking 10.1). The central administration may not understand the physical and social conditions across the country, and the incentive systems, such as wages for workers, may not lead to an efficient yield of products and services. Besides, farmers and other land managers may gain pleasure simply from being able to make their own decisions and from seeing the results of their work, which seems to motivate gardeners in rich countries to produce vegetables even though they would get them at a cheaper price and with much less work from a supermarket.

The other extreme, handing all decision-making to private landowners is equally problematic, but for other reasons. The resulting anarchy would not provide incentives for the decision-makers to consider others even though many would altruistically. In practice, countries and other administrative geographic units are balancing between these extremes and trying to control and incentivize people to manage land for the best of all. Often people have numerous legal or illegal yet tolerated rights on public lands, such as hunting, gathering, and the possibility of using a pasture for privately owned domestic animals. Correspondingly, in traditional societies with customary systems, private management is directed for the benefit of all villagers. Customary systems have decreased in importance in richer countries. Instead, state-organized systems based on financial incentives, legislation, and corresponding implementation become important. However, unexpected surprises could occur even with these tools (Weird thinking 13.2).

Weird thinking 13.2 How to incentivize those making decisions on land use?

I am sure that all fields of policies have perverse financial or legislative incentives that result in the opposite of what was intended. However, observing the perversity may take longer in forest policy, as changes in forests are slow and occur far from capital cities. Bad incentives may therefore be enforced for longer periods before being noticed and understood.

A typical case that has been explained to me on many occasions around the world is that land must be used more intensively than would otherwise be normal to secure private ownership of the area or the right to harvest the trees. Planting native species, letting natural succession forest an open area, or not harvesting an old-growth forest with high conservation value could initially be excellent solutions for both the landowner or land holder and at the national level, but if the forest is at risk of being taken away from the owner or holder for being "too green", in other words environmentally friendly, it is better to choose the option that is initially inferior for both parties. Forest with native species could be considered natural and therefore state owned. Natural succession in an open area could signal to those dealing with land tenure that the patch is not used and not needed. A privately owned old-growth forest could simply be too valuable to be logged from the perspective of authorities in charge of nature conservation.

It is not difficult to see the origin of these incentives. For example, protecting a forest is easiest done by simply forbidding its use, but the next

continued

continued

---

steps in the impact pathway should also be considered, so as to remove the incentive for the private forest owner to clear-cut in order to not lose tenure. Instead, the landowner should be positively incentivized to retain the forest with high conservation values as is. Obviously, this would require funds, which could be obtained by taxing those landowners that do not boost biodiversity in their forests, for example.

As legislative incentives are risky and may direct actions in the wrong direction and financial incentives are expensive and often bureaucratic, the action of people in forested landscapes is often influenced by others by providing technical assistance or guidance. Sometimes this really assists the locals in making decisions, but often the organizations providing the guidance try to make the locals act against their own interest. Governmental actors may try to convince the locals to act for the benefit of the whole nation; NGOs for their own goals, such as biodiversity conservation; and companies naturally wish to increase their profits. Therefore, using the positive word "guidance" is similarly misleading as the guidance itself.

Finally, I link these theoretical considerations with some practical patterns. The most benefits are generally spread from natural forests, while plantations spread less and croplands spread the least benefits. Larger units should therefore attempt to encourage smaller-level players to include more natural forests. This is happening, and the UN and European Union incentivize countries to incorporate large carbon storages in forests and national governments encourage individual forest owners to incorporate more natural forests and forests in general on their properties. However, the full picture is more confusing. Institutions, such as UNFCCC that organizes the annual climate gatherings, are very weak at the global or continental levels compared to national governments, and historical relicts and strong advocacy groups opposing any change are strong at the national level. Many countries have competing institutions. Those favouring forests are often more visible, but those focusing on croplands deliver significant subsidies, clearly beating the soft incentives for forestation. The wider picture is often conservative, so that land-use change is discouraged, and an area that has been agricultural is incentivized to remain as such despite natural forests typically being better for biodiversity, carbon, hydrology, and recreation. Therefore, my rough answer to the question "How much forest is needed?" in the title of this chapter is: quite a lot more than currently exists. If significant climate change mitigation is attempted with ecosystem management, much more forest area is required.

14

Forest Distribution

In the previous chapter, I did not consider large-scale variation in conditions where forests grow or could potentially grow. However, many things vary globally (Chapter 1), and optimizing land use based on observed empirical patterns is often done.[1] The number of tree and other plant and animal species, precipitation levels, and temperatures all decrease towards the poles. Climates become more continental and seasonal temperature variation increases away from the oceans. This increase is much more rapid against main wind directions, such as in eastern temperate Asia and North America. Excitingly, understanding biological processes allows modelling some of the patterns influencing global land-use optimization mechanistically. Because GPP and maintenance cost depend differently on temperatures (Figure 8.1), this inevitably influences growth and biomass accumulation.

When plotting actual data-based biomass accumulation curves for various humid climates, large differences are evident (Figure 14.1). Greatest old-growth forest biomasses are found when only small seasonal temperature variation is present, and for most of the year temperatures range around 10°C to 20°C, as in maritime temperate climates (Chapter 8). These forests are most valuable if mitigating climate change by conserving old-growth forests and their carbon stocks.

If the carbon storage of a forest does not change, increasing wood production should be beneficial for climate change mitigation, as carbon stored in wood products is away from the atmosphere, and wood can also substitute fossil fuels and materials causing significant carbon emissions (Chapter 11). Wood production is, of course, valuable even without considering climate change (Chapter 10). Biomass accumulation is fastest in warmer and less-seasonal climates (Figure 14.1). These would therefore be the best areas if forests are only kept for wood production.

Things become very interesting when we simultaneously consider carbon storage in forests and wood production. The above-ground biomass production in the extremely cold climate of Yakutsk is only 0.9 mg per year on a hectare of land with a 63-year rotation that maximizes production

Trees and Forests of the World. Markku Larjavaara, Oxford University Press. © Markku Larjavaara (2026).
DOI: 10.1093/9780197757109.003.0014

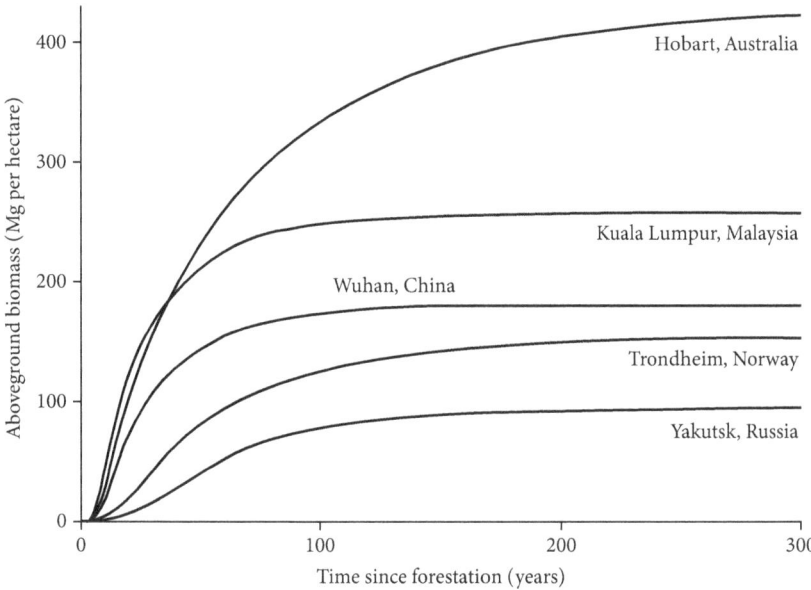

Figure 14.1 Biomass accumulation based on mathematical models calibrated based on a global dataset.[2] Biomass accumulation is fastest in the tropical lowland rainforest climate of Kuala Lumpur in peninsular Malaysia but stabilizes early in this region. Biomass accumulation is sustained for much longer in the maritime temperate climate of Hobart, Tasmania, and reaches a much higher level despite the slower start.

(Figure 10.1). Kuala Lumpur represents another extreme, with a 7.3-fold production per unit time relative to Yakutsk with a rotation of only 16 years. At the age of 300 years, corresponding roughly to maximal biomasses (Figure 14.1), the difference in above-ground biomasses between Kuala Lumpur and Yakutsk is only 2.7-fold. The climate of Yakutsk is much worse than that of Kuala Lumpur for both storing carbon in old-growth forest and for producing biomass, which corresponds roughly to wood production, but the difference is far greater in production than in storage. This means that when maximizing both production and storage, more production forests should be situated in the tropical climate and more carbon conservation forests in the cold climate.

The other three climates are intermediate based on the rotation age that maximizes above-ground biomass production and the production itself. However, the above-ground biomass at an age of 300 years is highest in the climate of Hobart and not in the Kuala Lumpur climate (Figure 14.1). When comparing maximal annual biomass accumulation and biomass at age of 300 years, the differences become evident. Only 39 years of maximal biomass accumulations are needed in Kuala Lumpur to reach an equivalent biomass as that at the age of 300 years. The corresponding number for Wuhan is 46 years, while for the climate in Hobart it is 75 years, 90 years for Trondheim, and 104 years for Yakutsk. Not only should cold climates be used more for carbon storage forests and the tropics for wood production forests, but the seasonality of temperatures matters in temperate climates; continental climates, such as in Wuhan, are more suitable for production, and maritime climates for storing carbon because of the temperature dependencies in physiology related to plant growth maintenance (Chapter 8).

Optimizing forestland allocation based on wood production and carbon storage so that lowland tropical forests would mainly be plantations would be disastrous for biodiversity when measured by global species numbers in various taxa, as tropical rainforests are by far the most species-rich terrestrial biome on Earth. If global extinction risks are included in the optimization, with focus on species groups with particularly steep latitudinal species gradients, such as trees, and enough weight placed on biodiversity, the result could be reversed so that instead of conserving boreal forests because of their slow growth relative to maximal biomass, tropical forests would be conserved for their biodiversity. However, biomes experience considerable variation in biodiversity, which is reflected in the mapping of biodiversity hotspots. Islands and mountainous regions are overrepresented as biodiversity hotspots,[3] and using them to both conserve carbon and biodiversity against global patterns would be wise. However, biodiversity can be defined in numerous ways (Chapter 12). The importance of the tropics decreases if the focus is on minimizing the absolute number of national extinctions, and this decrease is more pronounced if focus is placed on the proportion of all species going extinct nationally.

So far, I have not separately considered natural forests and plantations despite their differing appearances especially in the tropics, as natural forests there rarely recover from a recent disturbance or are rarely dominated by only a single tree species. In the boreal, even a local forester may occasionally find it difficult to distinguish planted monocultures from more natural successions. When considering optimal global forest distribution, for many of the benefits it matters greatly whether young forests are natural successions or plantations, especially in warmer climates. This is an active area of current research.[4] Species nativeness, species numbers, and management intensity all differently influence the benefits obtained from forests. However, here I begin comparing extreme cases: an intensively managed monoculture of an exotic species and a natural succession developing independently. These two extremes are similar when considering biomass accumulation.[2] Many natural secondary successions have little if any value for wood production, as most species may not be accepted by local forest industries. The very same reasons that make tropical secondary successions bad for wood production, such as the large number of tree species, make them good for biodiversity. For many other forest benefits, such as those related to hydrology and erosion, plantations, and natural secondary successions, are similar if the machinery used in managing the forests do not differ.[5]

Could other, more innovative hybrid approaches exist for producing wood efficiently without severely harming tropical biodiversity? Planting timber species and securing their survival and initial growth, but neglecting their subsequent management and allowing the infiltration of other species into the understorey or even into the main canopy, could be a solution far above the mean of secondary succession and plantation forest in terms of the value of wood produced and biodiversity conserved. These forests are often close to natural succession in biodiversity and to an intensively managed plantation in wood production.[6] They are commonly young and successional because trees have recently been harvested and the focus, also in the future, is on wood production. However, sometimes new forests are established without wood production objectives. Unfortunately, due to traditions, monocultures are established in regions that would naturally have a great number of species in natural successions. For example, young forests in China are often overmanaged due to such a historical burden, and a less intensive management style would be superior.[7]

When expanding the focus from optimal distribution within forests to all land use in areas that are humid enough to potentially be forests, we need to consider agricultural productivity. Agricultural plants are small and should therefore be similar to small trees that have a high GPP relative to maintenance cost and use the abundant surplus energy to either grow fast or develop harvestable produce. Simply based on this physiological perspective, agriculture should be most efficient per unit area in the warmest regions, where wood production also peaks. Then, with emphasis on carbon storage and agricultural production, the optimal pattern would be similar as with carbon storage and wood production. Agricultural production should be centred in the tropics and continental temperate climates. However, the theoretical predictions for agricultural yields do not match actual yields that are peaking in Western Europe, eastern United States, and China.[8] This is understandable, as these regions have invested decades into agricultural research and development, including breeding, and have highly optimized production systems in place with high inputs such as fertilizers and pesticides.[9]

Most agricultural plants are small, actually much smaller than trees in young forests. Most are also seasonal. The short life cycle allows cultivation for only part of the year (Weird thinking 1.3), and therefore some climates that are too dry for forests have efficient rain-fed agriculture during the wetter season. In contrast, the small size of most agricultural plants prevents them from growing deep roots and reaching water reserves deep underground that have accumulated during the rainy season or within the rainy spells of the dry season. This mechanism is more significant on sandy soils where water penetrates deep into the ground. Therefore, plantation trees may do better on sandy soils while seasonal agricultural plants fare better on clay soils.

When not focusing on the distribution of forests and croplands, but considering carbon in croplands, some potential approaches should be discussed. Increasing soil carbon in croplands could be one such solution, which has been widely discussed but unfortunately remains challenging.[10] However, dramatic impacts may be achieved at certain sites, such as previously waterlogged peatlands, which when drained and fertilized normally lose their carbon rapidly but where this process can be slowed down or even reversed by raising the water table.[11] Likewise, agroforestry clearly increases the above-ground biomass relative to a cropland where herbs are cultivated[12] and may increase food production in certain instances.[13] However, much greater results than attempting to find rare win–win cases could be achieved globally by searching for areas where a small decrease in food production could enable a large carbon gain thanks to trees and forests and vice versa. For example, large areas of the world, particularly in Africa, are used as pasture but with very small meat production per unit area,[14] and increasing forest biomass significantly would lead to only a small reduction in food production.[15]

The significant global variation in the suitability of land for carbon storage, biodiversity conservation, wood production, and agricultural output are partly already accounted for. Thanks to market mechanisms, some of this variation already influences current global land use. For example, agricultural land that could potentially offer higher net income from wood production may be converted into plantation. However, even this simple case in which the landowner optimizes their bottom line is often influenced by factors not directly related to the biological and labour-related processes on farms. When land tenure is uncertain longer investments are not favoured (Chapter 13), and governmental subsidies play a significant role in many regions. For reasoning that is incomprehensible to me, most rich countries have chosen to strongly subsidize their agricultural production but also forestry, leading to income transfers, often to wealthy landowners. Agricultural subsidies are even more incomprehensible when considering how the land use that provides the least benefits that spread beyond the landowner (Chapter 9) are subsidized more.

Biodiversity and carbon storage, which are normally not valued by markets, are poorly accounted for when optimizing cropland, production forest, and conservation forest distribution within given countries—and even more poorly on a global setting. Even though both the biodiversity and carbon storage of forests are discussed widely in global forums, practical action is restricted to declarations, pledges from individual countries, and projects managed by nongovernmental actors.

15

Global Change Impacts

When discussing in the previous two chapters how much forest is required and how it should be distributed globally, I have assumed that past growth conditions have been the same as they are currently and will be in the future. This is far from the truth because of global change, which I define here to include only anthropogenic changes and those that function via the atmosphere, therefore excluding phenomena such as land-use change (Chapter 9). Because forests influence global change factors (Chapter 11), potential positive feedback can cause a vicious cycle in which global change is reinforced by the changes it causes to forests.

The intensification of the greenhouse effect, which is the main cause of climate change, directly increases air temperatures because the proportion of outward radiation that is absorbed by greenhouse gases in the atmosphere is increased. Temperatures have already increased by more than a degree relative to the second half of the nineteenth century.[1] The warming has been faster at high latitudes, in winter, and during nights, which therefore reduces the spatial and temporal variation of temperatures. To understand the impact of changing temperatures on trees, we can focus separately on GPP and maintenance costs (Chapter 7). GPP increases with increasing temperature to a certain point and then collapses dramatically due to several mechanisms. The exact values for these depend on the species, its genotype, and acclimation based on temperatures of the past. Generally, the warmer the conditions in the evolutionary past and during an individual's lifetime, the warmer the optimal temperature. The climates of the past decades have shown that increasing temperatures decrease GPP in the lowland tropics but increase them elsewhere.[2] This indicates that under warming climates, GPP would increase in the boreal and normally also in the temperate biome. However, the situation may be different in lowland regions of the tropics and subtropics, where temperatures are already over the optimal and may be close to the temperatures leading to a dramatic decrease. As increasing the elevation by 100 m decreases temperatures around 0.5–1.0°C, the situation can be very different just a few hundred metres above sea level.

Trees and Forests of the World. Markku Larjavaara, Oxford University Press. © Markku Larjavaara (2026).
DOI: 10.1093/9780197757109.003.0015

The maintenance costs of trees increase with increasing temperatures. They are therefore expected to increase, as temperatures are expected to increase nearly everywhere. As GPP is more significant for small plants growing vigorously, the maximum growth rates are expected to increase with increasing temperatures except in the lowland tropics, in continental climates with hot summers, and in warmer distribution edges of tree species. For maximal biomasses, the ratio of GPP and maintenance cost matters, and this is expected to decrease except in the boreal biome.[3]

In many temperate forested regions,[4] and normally in the tropics,[5] tree diameter growth has been more sensitive to past precipitation variation than to past temperature variation. The enhancing greenhouse effect does not directly influence precipitation, but significant changes have been reported and larger ones are expected. With warmer temperatures, both transpiration from plants and evaporation from other surfaces increases, and a similar increase in precipitation takes place because atmospheric water cycles rapidly.[1] However, as most of the evapotranspiration originates from the oceans and most precipitation falls into the oceans, the continents could still experience a decrease in precipitation if the flow of water vapour from the oceans to the continents decreases significantly or the flow of water vapour from the continents to the oceans increases significantly. Most complex models generally predict increasing precipitation over land, while some areas may possibly experience decreasing precipitation, although the uncertainty is high.[1] However, warmer temperatures generally lead to increases in evapotranspiration, potentially reversing the impact to soil moisture that increasing precipitation alone would have. Potential impacts to soil moisture are further complicated by possibly decreased transpiration due to carbon dioxide fertilization (discussed below). In general, modelling studies suggest that soil moisture will more likely decrease,[6] which would then have negative consequences for GPP in areas where trees are unable to easily obtain water from deep soil layers. However, in many regions, especially in the equatorial tropics and boreal biome, soil can be saturated nearly to the surface, causing anaerobic conditions leading to challenges in anchorage and nutrient uptake (Sceptic's question 1.1). In these situations, decreasing soil moisture would naturally increase GPP. Climate change is altering the climate in numerous other ways than by just impacting temperatures and precipitation. For example, air humidity, photosynthetically active radiation, and wind speeds can also change significantly.

As plants take in carbon dioxide from the atmosphere during photosynthesis, increasing carbon dioxide production speeds up GPP. In fact, using abnormally high carbon dioxide concentrations has been a standard practice in the greenhouse cultivation of certain vegetables. In addition to the greenhouse effect, this carbon dioxide fertilization effect is therefore another important mechanism through which increasing carbon dioxide concentrations influence plants. Field experiments with carbon dioxide added to forests have indeed shown modest increases in the GPP of trees. Another potential way in which increasing carbon dioxide could influence forest ecosystems is related to the fact that to obtain the carbon dioxide levels required, trees would not need to keep their stomata so widely open, which would reduce transpiration.[7]

Another global change impact is eutrophication caused by nitrogen deposition. Combustion engines and numerous industrial processes emit nitrogen compounds that may end up in forest ecosystems. As nitrogen is an important element in leaves and is the main limiting nutrient in most forests outside the tropics and in some forests within the tropics, this nitrogen fertilization is likely to have significant impacts by boosting GPP. Unlike carbon dioxide and other greenhouse gases that spread evenly around the Earth within months, nitrogen fertilization is not evenly distributed. Much higher deposition occurs within or downwind of densely populated industrial regions of the world. Nitrogen deposition is still increasing in many poor countries, but decreasing in the wealthiest regions of the world.[8]

Forestry professionals are deeply concerned with the past and future global changes,[9] but we would very likely be even more concerned if the changes were occurring in opposite directions. Increasing temperatures, increasing precipitation, increasing carbon dioxide, and increasing nitrogen deposition are better than equivalent decreases. However, this does not mean that these changes would have a positive effect on the human use of natural forests and plantations because any change, even in a better direction, can be perceived as detrimental but perhaps not as catastrophic (Sceptic's question 15.1). This negative impact can be understood by envisioning an area with one-half having a given climate and the other half another type of climate. If these climates were suddenly switched, the overall average climate and its variation would remain the same yet the impacts to natural ecosystems, croplands, and plantations would mainly be terrible if the changes were sizable. Most agricultural plants are annuals, so even if the sudden switch of climates happened during the growing season, new crops would be rapidly imported from the other half of the area, and knowledge of how to cultivate them would build rapidly in modern societies. Therefore, much of the production would have been recovered within a few years. In the case of timber plantations or agroforestry, the long rotation related to trees would mean that recovery takes much longer if all the trees were killed in the sudden switch. However, current global change is slow and subtle relative to the above-described sudden change. Despite this, similar difficulties but of smaller magnitude may arise. These problems could be serious to natural ecosystems, their biodiversity, and especially to more traditional agricultural communities. However, the relative importance of agriculture and forestry is decreasing, and the most important impacts of climate change are not those influencing natural and human-modified terrestrial ecosystems but the oceans, particularly sea level rise.

Sceptic's question 15.1 Will climate change first kill all the trees and then all of us?

I have kept the paper on which I handwrote the morning talk that I gave to my secondary school in 1990. It contains many naiveties, but I agree with my 15-year-old self on climate change being the most serious environmental threat that we face and its importance relative to other global challenges. Perhaps this is not surprising, as the scientific basis was well known back then, and even I knew quite a lot about it. However, current media discussion based on the volume and tone is perhaps two orders of magnitude more intense than it was in those days. I was significantly more concerned about climate change than most people around me, but now the situation has reversed. This is even more surprising when bearing in mind that people are constantly becoming less and less dependent on the climate thanks to improving technologies and the lower importance of agriculture in economies.

What is the right level of concern? Research is surprisingly bad at answering questions such as this. For example, you might first think that taking a large sample of all relevant scientific publications would cancel out biases. This would be true if there was no general bias that unrealistically requires not only fully objective data collection, analysis,

continued

continued

---

and reporting, but also initial selection of the research topics free of values and interests of the researchers, their supervisors, or of their donors funding the research. Paradoxically, one could even argue that biases are a greater problem in reviews or meta-analyses, as readers tend to assume greater objectivity than they would with individual studies.

Another layer of potential bias in the media discussion comes from journalists selecting the research that they cover. Both scientists and journalists are then prone to information cascades, where already biased information is biased further (Weird thinking 11.1), normally unintentionally. Empirical research is the best way of understanding the world, and it is particularly valuable in removing the information cascades that lead our understanding astray. However, as empirical research is itself biased and scientist adopt methods from other researchers, active inspection by numerous people is needed, and educated guesses are required regarding the potential biases that me, other scientists, and all others suffer from.

Many people are probably saying that scientists need to exaggerate the impacts of climate change because too little is being done despite current exaggerations, and many more are thinking this. I agree that far too little is being done, but I strongly disagree that scientists or journalists should intentionally bias results to achieve a political goal (Chapter 16). Instead, even if we can never reach the goal, we should nonetheless always aim for the truth.

I now focus again only on maximal biomasses; I do not consider growth in successional forests. In addition, I assume a sudden change in climate, carbon dioxide, and nitrogen fertilization. If the new climate, with carbon dioxide and nitrogen fertilization effects, only enables a smaller biomass than the previous climate, the speed of the change in forest structure and species composition depend on the initial structure and species. If the stand is relatively evenly sized, and as the trees cannot become smaller, the large original trees are unable to invest sufficiently in their protection due to a negative energetic budget and are killed by some disturbance. However, if the original stand structure is uneven, the largest trees may be killed and smaller trees from beneath become the new large trees. However, they will never reach the size of the original large trees, as the biomass of the new climate was assumed to decrease. This replacement may take plenty of time if the replacing trees are small at the onset of the sudden change, and even longer if they must migrate from other parts of the same landscape or even from other climate zones. In an extreme case, if the species adapted to the new conditions must first evolve, the transitional phase may take millions of years, if assuming unrealistically that conditions remain constant after the sudden change.

In an equivalent situation but with increasing biomass due to the sudden change, the patterns could vary dramatically depending on whether the dominant species remain dominant. If yes, then the large trees can simply grow even larger. If not, then the original large trees may persist if the conditions after the sudden change allow, or they will die due to the changes and the biomass will bounce back.

Interestingly, as theoretically discussed above, changes in maximal biomass may occur in both directions via disturbances (Figure 15.1). Therefore, it would be useful to distinguish between increasing disturbances of the transitional period, related to the original species not being suitable for the new conditions as described above, and permanently changed disturbance regimes. However, even transitional changes could lead to a long-term or even permanent decrease in maximal biomass if the ecosystem tips into a new steady state sustained by frequent fires, for example. It is difficult to say anything of potential permanent changes in disturbance regimes that are not triggered by the transitional stage. Claims could be made that increasing extreme wind speeds could increase wind disturbances, but such reasoning is flawed because trees could simply build trunks that are able to sustain themselves in the new conditions (Sceptic's question 3.1).

Mechanistic modelling is one way to study the impacts of global change. Experiments with increased carbon dioxide or higher temperature offer another potential approach. The third option for understanding global change impacts is by studying changes of the past decades and assuming that the empirical trend will continue without considering the mechanisms causing the changes. All these three approaches have significant problems, and therefore our understanding of the future remains hazy. Both the modelling and empirical trend approaches suffer from a fundamental problem, as the usual scientific approach with deriving hypotheses from theories and testing them cannot be achieved now but instead the support or rejection comes only decades after. The modelling approach is challenging because

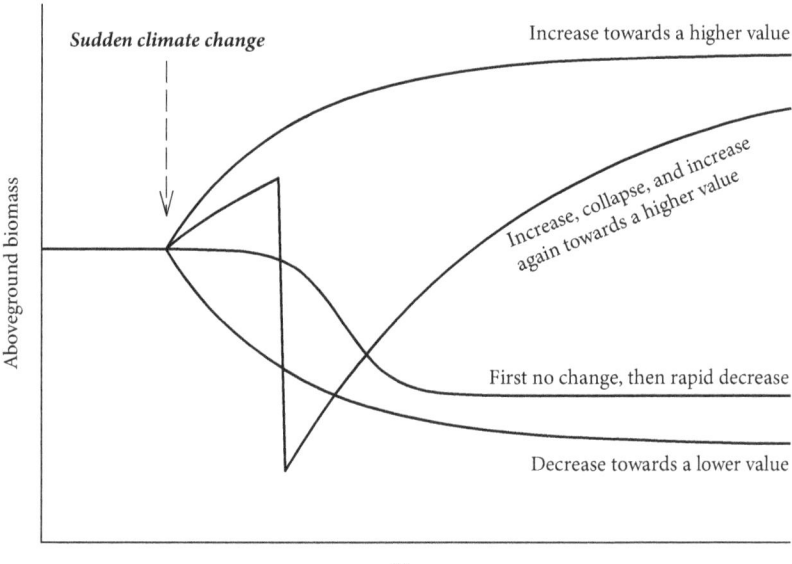

Figure 15.1 Hypothetical paths of above-ground biomass developments after a sudden and permanent climate change. These and nearly all other paths are possible, and only an increase that is more rapid than the rate at which trees are growing is impossible. An initial development is not a guarantee of future development in the same direction. The real situation is even more complicated, as several global change factors change gradually and simultaneously.

the ecosystems are complex, and realistically describing even the observed spatial and temporal patterns has been difficult. The experimental approach is hampered by a scale issue. Trees are large, but increasing carbon dioxide or raising temperatures can only be done at small scales.

The empirical trend approach is suffering from many additional challenges. First, understanding what has happened is challenging even with high-quality data.[10] The challenges in ensuring data quality in long-term series are more demanding than most people who have not participated in such research could possibly realize. Errors may then lead to dramatic trends published in authoritative journals, even if such trends do not actually exist, or if they exist only in the experimental plots. This could, for example, happen if the research towers act as lightning rods, decrease lightning-caused mortality, and therefore increase biomass. Second, even if we understand the past trends, this may not describe the future. The past trend could be caused by fluctuations not induced by global change (Sceptic's question 1.2), or the observed changes could be related to the transitional phase (Figure 15.1). The transition is more likely to involve a drop rather than a temporary increase in biomass, and the changing conditions could be lethal to many original species, leading to trees dying either individually or in larger disturbances (Sceptic's question 15.2). Therefore, the slowly increasing biomass densities in tropical lowland rainforests[11] would indicate that the trend will continue and is unlikely to be reversed soon. However, as the causes of the increase are uncertain, even a reversal is possible. This could happen if the recent increases have been caused by carbon dioxide fertilization and increasing temperatures would have an opposite effect that is initially smaller but could become larger, or if the increasing aridity causes more aridity and deforestation via positive feedback. Correctly attributing subtle changes in biomass to global change impacts is challenging in other biomes due to large-scale disturbances following succession (Sceptic's question 1.2).

Sceptic's question 15.2 Are natural disturbances increasing?

Yes, they are, as the new climate may be energetically less conducive for supporting the existing biomass, or even if maximal biomass is increasing, some of the tree species may not be able to cope, will weaken, and can eventually die. Using a wide definition of "disturbance", is part of most imaginable global change scenarios (Figure 15.1). However, this is not the typical reasoning behind claims stating that disturbances are increasing.

Normal thinking begins with observing or expecting an increasing intensity of a disturbance agent, such as wind, and then assuming that more trees are snapped or uprooted because of this increase. However, disturbance agents vary in how trees can sense and respond to them. The agents can be continuous, such as wind, or discreate events of variable frequencies like fire. Trees sense changes in the continuous agents, but this is impossible in the case of infrequent discrete events. Also, the potential response rates of ecosystems vary greatly from the induced defences against herbivores in minutes, to acclimation against winds in months, and to fire responses where changes in species composition take decades or longer if migration is needed. Trees should not have

continued

continued

trouble in acclimating to stronger winds (Sceptic's question 3.1). However, trees cannot acclimate to more frequent forest fires because there are no fires during the decades in between fatal blazes, and therefore evolution might not even have developed ways for trees to acclimate. Pests and pathogens are somewhere in between, depending on how their populations fluctuate and how trees are able to prepare against their attacks. If a pest population increases also between outbreaks, then trees may be able to acclimate to the pest's higher population densities.

Even when trees cannot sense and respond to the potentially changing disturbance agents, the agent may not be increasing despite most forest scientists and especially the public thinking otherwise due to biased media reporting. For example, the globally burned area has decreased in the past decades.[12] Even when not relying on media reports, human memory is bad at noticing changes over decades. Our mind simplifies the messy observations from the distant past into simple patterns.[13] For example, people in seasonally dry tropics are convinced, based on their memories, that decades ago dry seasons were rainless, wet seasons were very rainy, and current rains during the dry season or dry spells in the middle of the rainy season are caused by climate change. However, when analysing climate statistics, the evidence typically does not support these biased perceptions, potentially stemming from our tendency to remember the rainiest days of the rainy season more vividly and to compare them with a typical rainy season day in the current season.

16

Forests and Forest Science in the Future

Describing the past changes in the world's forests (Chapter 9) in a succinct way is challenging. Predicting future changes is many times more difficult but less stressful, because it is not possible to prove me wrong during my lifetime if the predictions I make apply to a longer period.

How would a forest historian in 2125 write about land-use history and forest area trends of the previous century? It would be too simplistic to naively assume that a similar forest transition pattern as that in European countries would hold for countries that are in the rapid deforestation phase or that have not yet even entered this phase. Global trade is one reason why countries that are last in the queue are unable to follow the path trampled by nations that industrialized first. The natural market-driven pattern with easily transportable agricultural products, such as grains, is that their production moves to poorer countries. In the case of the poorest countries, poorer countries do not exist, and labour-intensive agriculture cannot move to cheaper countries. The natural patterns dictated by free economy have been confounded by agricultural subsidies, which have dramatically influenced land-use patterns in wealthier countries, and indirectly in the remaining regions as they have lowered crop prices in global markets, probably reducing tropical deforestation significantly. This has been harmful from the carbon storage and wood production perspective (Chapter 14) but beneficial for global biodiversity.

Much depends on the future of agricultural subsidies. Another area of great uncertainty is land usage for mitigating climate change (Chapter 11). If this is seriously attempted, the impacts may rise to at least the same magnitude as those of current plantation forestry. More likely is the business-as-usual future scenario and a continuation of the processes that have already begun, including the continued decrease in forest area at a decelerating rate and increasing emphasis placed on carbon storage and biodiversity, with significant conservation attempts aimed at globally and nationally endangered species.

Trees and Forests of the World. Markku Larjavaara, Oxford University Press. © Markku Larjavaara (2026).
DOI: 10.1093/9780197757109.003.0016

Global change will impact forests in the upcoming century more than it has in the past century (Chapter 15). The growth rates of trees are likely increasing in general, but old-growth biomass densities could very well decrease in large areas, particularly in lowland tropics. Despite this, it does not seem likely that a strong, vicious cycle would speed up the increase of carbon dioxide in the atmosphere, as boreal forests may offer valuable negative feedback with more carbon stored. Numerous other positive ecosystem feedback have been described in addition to the decrease of tropical tree biomass, and some of these could be significant. Fortunately, the number of negative feedback is also large. Although understanding global change impacts is challenging, understanding the often many times more significant local drivers of deforestation, forestation, and other ecosystem management is even more challenging and more important in most cases. We forest scientists and others interested in changes in the world's forests often underestimate the impact of local processes and overestimate global ones. This seems to be caused by two mechanisms. First, global change impacts are much easier to quantify, especially for the future. Understanding local processes, even in the past, requires tedious detective work.[1] Second, people prefer not to blame actors that are physically close for the negative changes especially if they belong to historically exploited peoples, but rather it is easier to criticize the general global population for the problems.[2]

Future forests that would be optimal for humankind differ drastically from forests that we will get with business-as-usual. I have explained how central the time frame based on which the optimization is performed is in forest-related decision-making. Forests are often conserved or established when planning is conducted far into the future, and these differing optimization periods are most elegantly accounted for with interest rates (Figure 9.1). However, the changes in forest ecosystems towards more biomass are so slow that we actually should not optimize land use solely based on current values and preferences, but that we should estimate their change in the upcoming decades and centuries. As people's values have generally been changing towards valuing soft benefits more, such as biodiversity, and wood production less, current policies should be greener than based on current values. There are several additional reasons why policies for the common benefit should be even greener. In democratic societies, those eligible to vote, and typically even more those who do vote, are older than the overall population while younger people have greener views, again skewing the optimal towards greener policies relative to what politicians are actually advancing.

Many practices simply remain stuck in the past, and this problem could worsen in natural resource management, which perhaps has more conservative people involved than other fields do. Unrelated to the time lag, another layer pushing the overall optimal even further towards greener policies comes from the divergence of individual and general benefits (Figure 13.1). Because of all these reasons, we should favour greener initiatives in national environmental policies to incentivize softer forest management and to reduce deforestation. When the optimum for all global people is estimated, the policies should be even greener. I have written "however" over 200 times in this book, but this time there is no however. Instead, all the unaccounted mechanisms point in the same direction. The great tragedy when examining global land-use history is that decisions have been made by focusing on a single product or a few products while not paying enough attention to services. We are still on the same incorrect path leading to significant deforestation.

The current global forest distribution is far from optimal. For historical reasons, people live in cities and rural areas where food production is most efficient. This is imperfect for several reasons. First, the buildings, roads, and other infrastructure use land that would also be valuable for crop production. Second, these regions are not ideal for recreation. However, there is not much that can be done to incentivize people to move to regions with less productive land, even in a decadal time frame. Forests for recreation can be brought close to cities within this time frame, as has been done in most middle- and high-income countries. Unfortunately, many of these forests resemble timber plantations or intensively managed parks, even though more natural forests would better serve especially those recreational users who value biodiversity.

Knowing what exactly we want is difficult because of the biased nature of available research, but clearly we need greener forest policies. Seeing the route to this destination is unfortunately even more difficult. Even when we do see the way forward, we are often too optimistic when considering foreign countries without understanding their societies. For example, many of us in Finland believe that breaking the status quo and forcing Finnish landowners to change the management of their peat soil croplands is difficult or impossible, even though doing so could result in huge avoided carbon emissions. However, the same Finnish people consider it a viable option to use money to persuade peatland managers in the Indonesian part of Borneo to avoid emissions, even though the challenges related to ethnic and religious tensions, land tenure, carbon accounting, and designing compensation mechanisms are immense compared to the Finnish case.[3]

The softer path that directs forest managers to act towards the common benefit is based on changing the ethics and values of people. Even the rise of European powers over the past 600 years has been linked to a shift from a focus on self-interest and extended family ties to a greater emphasis on national goals.[4] In wealthy countries, more climate change mitigation has possibly happened by people simply being willing to contribute by driving electric vehicles and installing solar panels on their roofs than based on planned policies and regulations coming from above. Unfortunately, such moral-based incentives work best when peers can witness the action. Therefore, these mechanisms have the potential to make people grow increasing numbers of flowers for pollinators in front of their domiciles, but they do not incentivize increasing the carbon storage of faraway forests that nobody sees.

The more authoritarian path, which is definitely working, is the usual top-down dictation of legislation and economic incentives that direct decision-makers regarding forest area or forest management. Making people exchange their personal freedoms to better the lives of all citizens is one of the main reasons why nations exist. We require stronger national governments that make environmental legislation stricter, and to create the best circumstances for all countries we need global processes in which countries relinquish their freedom to make environmental decisions.

Information and understanding on how to respond to the negative changes caused by global change and local actors is critical for designing optimal strategies. As usual in policy design, opposing advocacy groups try to influence forest policymaking. As there is a clear trend towards greener policies, groups opposing them try to keep the status quo by claiming that there is no conflict or alternatively by trying to divert the attention of people to some minor issues and to keep discussions focused on these issues as long as possible. The argumentation biases made by groups that are often associated with private landowners or users of wood, other forest products, or alternative land uses are relatively easily identified by most followers of the debate. Biases in the narrative of the opposing force—environmental organizations—have been less easy to detect because much of the argumentation has been based on the cryptic term "biodiversity" (Sceptic's question 12.1), and the groups have argued that they act for humankind. In reality, these organizations should often be considered advocacy groups for nature lovers only. Groups on both sides of the disagreement only rarely describe the conflict of interests (Chapter 13), and instead the green side argues that everyone benefits from more biodiversity while the opponents claim a vibrant economy to benefit all.

As the advocacy groups are biased almost by definition and definitely in practice, forest science must be the basis of our response. And we need all kinds of forest science. Not just the traditional inventories, experiments, and modelling, but also softer methods that are perhaps not currently acceptable in forest sciences, such as wandering and pondering in the forest.

In 2125, when historians reflect back on forest science in the twenty-first century, they will be able to identify large biases in many areas of the field. They will wonder how scientists in the first quarter of the twenty-first century found so much support for their favourite hypotheses and how these studies were accepted even in the top scientific journals, as they later turned out to be more multifaceted questions. For example, more diverse forests now appear to increase productivity and to one-sidedly also be better in every other aspect, and global change causes almost solely negative impacts.

Journalists and the public are deeply convinced that the research is correct. Peer review, in particular, builds trust and a meta-analysis or review of numerous peer-reviewed articles even more so. In reality, peer-review is a quick procedure typically taking less than one per cent of the person hours of the research process itself. Biases in science are more dangerous than in discussions with advocacy groups, as many of those listening to researchers incorrectly assume science to be mainly free of biases.

The first category of biases stems from methodological issues. Scientists may end up favouring simple explanations, even when alternative but more complex ones are more likely. For example, since past and future climatic changes are well known, it is tempting to attribute too many changes in forests to climate.

The biases in the second category originate from scientists aiming for personal benefit by biasing the message of their research. In an extreme example, a researcher takes cash from a profit-making company and in return gives a more positive report of a product manufactured by the company in a scientific publication. In milder cases, a conservation biologist could inflate the value of a forest to get their recreational area protected (Sceptic's question 12.2), a forestry student could avoid certain thesis topics to have better chances of being employed by forest industries, a climatologist could report only part of their dataset to make more friends in an annual workshop, or a forest ecologist (me) could be more willing to conduct field work in scenic landscapes with larger trees (Yunnan, China) than in other regions. Offering and accepting cash is clearly immoral but the other cases are not for most people.

The biases in the third category do not originate from seeking personal benefit but from scientists reasoning that they have to bias the message to influence the policymakers or other people towards the direction that they consider the right one for all humans. This seems to be the normal way of thinking for biodiversity (Sceptic's question 12.1) and climate change scientists (Sceptic's question 15.1). In my opinion, this should be considered immoral and is also unwise, at least in the long term. When actors sense the exaggeration, the risks of developing a counterforce and polarization are real, and even more dangerously a general loss of confidence in science may ensue. We should accept and be open about how many people, most perhaps in the short term, are on the losing side concerning the changes we propose. Nevertheless, these changes must be made anyway. Stories in mass media of how an individual land manager does not understand their own best and does not sequester carbon in their forest and will therefore suffer from the impacts of climate change are seriously misleading. It is not difficult for a land manager to understand how the fruits of their own actions are diluted among all world citizens and how free-riding and letting others do the mitigation is better for them personally.

Most taxpayers do not want scientists to intentionally bias the message. By not biasing intentionally we scientists have less power, as our messages are more ambiguous, but we should nevertheless aim for this. We should concurrently understand how difficult it is to avoid biasing the message, even in research that is not directly policy related.[5] This is the case even when nothing abnormal is being done. We can follow the usual procedures within the selected methods. However, some individuals are forced to make many subjective decisions prone to biases. To decide what to study is a subjective decision of the researcher, of their supervisor, and of the research funder. All these agents can have expectations of the potential results and may not choose to study or fund the topic if it does not seem to support the favourite theories or yield results that are socially acceptable. As a budding doctoral student, you know that you will graduate in the allowed time frame if you obtain statistically significant results and support for the tested theory, but your results will never be published if you randomly select the study sites and choose the most appropriate statistical method. Instead, would you study a forest that seems to show trends supporting the theory and test dozens of statistical approaches, only to report the one that yields significant results? Psychological research has been damaged by the "replication crisis".[6] In forest sciences, differences in the results can normally be attributed to variation among sites even though similar biases are probably underlying. Similar subjective decisions also bias the dissemination of research results. Whether a scientist decides to push for a press release by their institution, whether a journalist chooses to publish the news in their newspaper, and whether a reader decides to read it, all potentially bias the picture that the reader gets from the original research.

Science is by far the best way to explain the world, its forests and trees, but we need to accept that all science is inevitably more or less biased, and we should continuously fight for it to be the latter. We scientists should be intellectually honest, be courageous, and speak up when we see a consensus forming away from reality. We should try to forget the social dynamics and policy implications and simply aim for the truth.

Epilogue

Examining whether there are common patterns in the flawed reasoning presented in the Sceptic's questions, Weird thinking text boxes, and elsewhere in the book may offer valuable insights into the mechanisms by which human reasoning related to trees and forests often goes astray.

Reasoning can be narrowly focused, with indirect effects or subsequent steps in the impact pathway that might alter the practical implications often overlooked. Examples of this excessively narrow focus include the often-discussed leakage in climate change mitigation (Chapter 11) or the perverse economic or legislative incentives that result in local actors doing the opposite of what was intended (Weird thinking 13.2).

What is common for all the established ways of thinking that I criticize in this book is that confirmation bias plays a part in the development of the incorrect thinking. Many have sought arguments or research settings from which the favourite theories gain support instead of objectively comparing the pros and cons. A typical example is the overly positive attitude relative to agroforestry (Sceptic's question 13.1). Alarm bells should ring when someone lectures one-sidedly about the benefits of democracy (Weird thinking 9.2), dangers of climate change (Sceptic's question 15.1), or how trees help each other (Weird thinking 6.3). This kind of one-sidedness not only obscures the truth but can also contribute to dangerous polarization. Instead, we should aim to consider both the pros and the cons.

However, making meaningful comparisons can be difficult. Sometimes a comparison of sorts exists, but the reference is unclear (Sceptics question 11.2). However, this kind of vagueness is not that common in scientific literature, but instead unfair comparisons are. Most of us are good at listing pros and cons but figuring out what to compare is more challenging. Intuitively choosing what is assumed to be constant in wood density studies has directed a large group of researchers to make incoherent comparisons (Weird thinking 6.1). Similarly, reasoning the pros and cons of buttresses (Sceptic's question 5.1), multiple stems for an individual (Sceptic's question 7.2), and large leaf size (Sceptic's question 2.2) are all susceptible to this same cognitive dead end. Likewise, a comparison of tree mortality risk with the old and future climate has been done but is pointless in many cases (Sceptic's question 15.2). The same trap is also common in forest management in comparisons of rotational forestry and continuous cover forestry (Sceptic's question 10.2) and in judging whether agroforestry should be promoted (Sceptic's question 13.1).

Finally, many flawed conclusions arise from our inability to recognize when things are already at an appropriate level or in equilibrium and apply that understanding in our reasoning. When a major football tournament has been held, I have organized a betting ring for the kids and teenagers in my extended family. My objective has not been to win back their weekly allowances but to make them understand how they should not only consider which country will win but also how the other betters will bet because that influences the winning odds in the zero-sum game. Our mind easily drifts to simply focusing on the first layer and contemplating which country will win the game, or, when discussing shares in stock exchange, which company is good, without paying enough attention to the invisible hand that has already adjusted the betting odds and share prices to the right level. Similarly, evolution often functions as the invisible hand. Tree genes are already optimal for the conditions in which they have evolved, making breeding faster growth in those same conditions challenging (Sceptic's question 10.1); trunks are of ideal diameter once trees have had enough time to acclimate (Sceptic's question 3.1); and explaining how evergreens and deciduous trees must be approximately equally fit in the landscapes in which they coexist (Sceptic's question 2.1). Analogously, in traditional cultures, another invisible hand adjusted human population density at its maximum level and starving was common in all environments (Sceptic's question 9.2).

The first two causes of faulty reasoning are well-known and widely discussed. Naturally, it is wise to consider impacts one step further and weigh pros and cons in a balanced way. The latter two causes, improper framing of comparisons and neglecting equilibrium, are more intriguing and deserve greater attention.

About the Author

Markku Larjavaara is a forester and associate professor at the Department of Forest Sciences of the University of Helsinki, Finland. He was born in 1975 and grew up in Helsinki but spent nearly all his school holidays in a remote district in southeastern Finland enjoying slow solitary walks in forests and gazing at the trees and other wildlife. Markku has worked in both development aid and research studying tree structure and growth, ecosystem carbon, and forest management. He has slept 400 nights in a tent, lived on five continents, and watched the world's forests and land uses during his long over-land journeys that have crossed the land borders of countries in which over six billion people live.

References

Prelims

1. Park, M., Leahey, E., and Funk, R.J. (2023). Papers and patents are becoming less disruptive over time. Nature *613*, 138–144.
2. Kahneman, D. (2011). Thinking, fast and slow (Macmillan).

Chapter 1

1. Bar-On, Y.M., Phillips, R., and Milo, R. (2018). The biomass distribution on Earth. Proceedings of the National Academy of Sciences *115*, 6506–6511.
2. Chao, S. (2012). Forest peoples: numbers across the world (Forest Peoples Programme Moreton-in-Marsh).
3. Behrenfeld, M.J. (2014). Climate-mediated dance of the plankton. Nature Climate Change *4*, 880–887.
4. Verdú, M. (2002). Age at maturity and diversification in woody angiosperms. Evolution *56*, 1352–1361.
5. Piovesan, G., and Biondi, F. (2021). On tree longevity. New Phytologist *231*, 1318–1337.
6. Wellman, C.H. (2010). The invasion of the land by plants: when and where? New Phytologist *188*, 306–309.
7. Stein, W.E., Mannolini, F., Hernick, L.V., Landing, E., and Berry, C.M. (2007). Giant cladoxylopsid trees resolve the enigma of the Earth's earliest forest stumps at Gilboa. Nature *446*, 904–907.
8. Niklas, K.J. (1997). The evolutionary biology of plants (University of Chicago Press).
9. Biswas, C., and Johri, B.M. (2013). The gymnosperms (Springer).
10. Cronk, Q.C., and Forest, F. (2017). The evolution of angiosperm trees: from palaeobotany to genomics. In Comparative and evolutionary genomics of angiosperm trees (Springer).
11. Jud, N.A., D'Emic, M.D., Williams, S.A., Mathews, J.C., Tremaine, K.M., and Bhattacharya, J. (2018). A new fossil assemblage shows that large angiosperm trees grew in North America by the Turonian (Late Cretaceous). Science Advances *4*, eaar8568.
12. FAO (2024). The state of the world's forests 2024.
13. Ghazoul, J., and Medina, L. (2024). Deforestation and land clearing. In Encyclopedia of biodiversity, third edition, Volume 1–7 (Elsevier).
14. Feulner, G., Rahmstorf, S., Levermann, A., and Volkwardt, S. (2013). On the origin of the surface air temperature difference between the hemispheres in Earth's present-day climate. Journal of Climate *26*, 7136–7150.

15. Larjavaara, M. (2014). The world's tallest trees grow in thermally similar climates. New Phytologist *202*, 344–349.

16. Vinod, N., Slot, M., McGregor, I.R., Ordway, E.M., Smith, M.N., Taylor, T.C., Sack, L., Buckley, T.N., and Anderson-Teixeira, K.J. (2023). Thermal sensitivity across forest vertical profiles: patterns, mechanisms, and ecological implications. New Phytologist *237*, 22–47.

17. Konrad, W., Katul, G., and Roth-Nebelsick, A. (2021). Leaf temperature and its dependence on atmospheric CO_2 and leaf size. Geological Journal *56*, 866–885.

18. Gauthier, S., Bernier, P., Kuuluvainen, T., Shvidenko, A.Z., and Schepaschenko, D.G. (2015). Boreal forest health and global change. Science *349*, 819–822.

19. Dial, R.J., Maher, C.T., Hewitt, R.E., and Sullivan, P.F. (2022). Sufficient conditions for rapid range expansion of a boreal conifer. Nature *608*, 546–551.

20. Cao, X., Tian, F., Herzschuh, U., Ni, J., Xu, Q., Li, W., Zhang, Y., Luo, M., and Chen, F. (2022). Human activities have reduced plant diversity in eastern China over the last two millennia. Global Change Biology *28*, 4962–4976.

21. Loubota Panzou, G.J., Fayolle, A., Jucker, T., Phillips, O.L., Bohlman, S., Banin, L.F., Lewis, S.L., Affum-Baffoe, K., Alves, L.F., and Antin, C. (2021). Pantropical variability in tree crown allometry. Global Ecology and Biogeography *30*, 459–475.

22. Muscarella, R., Emilio, T., Phillips, O.L., Lewis, S.L., Slik, F., Baker, W.J., Couvreur, T.L., Eiserhardt, W.L., Svenning, J.C., and Affum-Baffoe, K. (2020). The global abundance of tree palms. Global Ecology and Biogeography *29*, 1495–1514.

23. Fine, P.V., and Ree, R.H. (2006). Evidence for a time-integrated species-area effect on the latitudinal gradient in tree diversity. The American Naturalist *168*, 796–804.

24. Cazzolla Gatti, R., Reich, P.B., Gamarra, J.G., Crowther, T., Hui, C., Morera, A., Bastin, J.-F., De-Miguel, S., Nabuurs, G.-J., Svenning, J.-C., et al. (2022). The number of tree species on Earth. Proceedings of the National Academy of Sciences, *119* e2115329119.

25. Jia, S., Wang, X., Yuan, Z., Lin, F., Ye, J., Lin, G., Hao, Z., and Bagchi, R. (2020). Tree species traits affect which natural enemies drive the Janzen-Connell effect in a temperate forest. Nature Communications *11*, 286. 10.1038/s41467-019-14140-y.

26. Hülsmann, L., Chisholm, R.A., Comita, L., Visser, M.D., de Souza Leite, M., Aguilar, S., Anderson-Teixeira, K.J., Bourg, N.A., Brockelman, W.Y., and Bunyavejchewin, S. (2024). Latitudinal patterns in stabilizing density dependence of forest communities. Nature *627*, 564–571.

27. Pontarp, M., Bunnefeld, L., Cabral, J.S., Etienne, R.S., Fritz, S.A., Gillespie, R., Graham, C.H., Hagen, O., Hartig, F., and Huang, S. (2019). The latitudinal diversity gradient: novel understanding through mechanistic eco-evolutionary models. Trends in Ecology & Evolution *34*, 211–223.

28. Wiegand, T., Wang, X., Fischer, S.M., Kraft, N.J., Bourg, N.A., Brockelman, W.Y., Cao, G., Cao, M., Chanthorn, W., and Chu, C. (2025). Latitudinal scaling of aggregation with abundance and coexistence in forests. Nature. *640*, 967–973

29. Fajardo, A., McIntire, E.J., and Olson, M.E. (2019). When short stature is an asset in trees. Trends in Ecology & Evolution *34*, 193–199.

30. Luo, A., Xu, X., Liu, Y., Li, Y., Su, X., Li, Y., Lyu, T., Dimitrov, D., Larjavaara, M., and Peng, S. (2023). Spatio-temporal patterns in the woodiness of flowering plants. Global Ecology and Biogeography 32, 384–396.

31. Schuur, E.A.G. (2003). Productivity and global climate revisited: The sensitivity of tropical forest growth to precipitation. Ecology 84, 1165–1170.

32. Bucci, S.J., Goldstein, G., Scholz, F.G., and Meinzer, F.C. (2016). Physiological significance of hydraulic segmentation, nocturnal transpiration and capacitance in tropical trees: paradigms revisited. In Tropical tree physiology: Adaptations and responses in a changing environment (Springer).

33. Resco de Dios, V., Chowdhury, F.I., Granda, E., Yao, Y., and Tissue, D.T. (2019). Assessing the potential functions of nocturnal stomatal conductance in C3 and C4 plants. New Phytologist 223, 1696–1706.

34. Minkkinen, K., Ojanen, P., Koskinen, M., and Penttilä, T. (2020). Nitrous oxide emissions of undrained, forestry-drained, and rewetted boreal peatlands. Forest Ecology and Management 478, 118494.

35. Guedes, B.S., Olsson, B.A., Sitoe, A.A., and Egnell, G. (2018). Net primary production in plantations of Pinus taeda and Eucalyptus cloeziana compared with a mountain miombo woodland in Mozambique. Global Ecology and Conservation 15, e00414.

36. Fan, Y., Li, H., and Miguez-Macho, G. (2013). Global patterns of groundwater table depth. Science 339, 940–943.

37. McDonnell, M.J. (2011). The history of urban ecology: An ecologist's perspective. In urban ecology: Patterns, processes, and applications (Oxford University Press).

38. Pickett, S.T.A., and White, P.S. (1985). The ecology of natural disturbance and patch dynamics (Academic Press).

39. Cheng, Z., Aakala, T., Ji, C., and Larjavaara, M. (2025). Disturbance dynamics and its effects on carbon in human-impacted mountain forests in northwestern Yunnan, China. Ecology and Evolution 15, e72165.

40. Jackson, R.B., Lajtha, K., Crow, S.E., Hugelius, G., Kramer, M.G., and Piñeiro, G. (2017). The ecology of soil carbon: pools, vulnerabilities, and biotic and abiotic controls. Annual Review of Ecology, Evolution, and Systematics 48, 419–445.

41. Wallenius, T. (2011). Major decline in fires in coniferous forests: reconstructing the phenomenon and seeking for the cause. Silva Fennica 45, 139–155.

42. Palviainen, M., Lauren, A., Pumpanen, J., Bergeron, Y., Bond-Lamberty, B., Larjavaara, M., Kashian, D.M., Koster, K., Prokushkin, A., Chen, H.Y.H., et al. (2020). Decadal-scale recovery of carbon stocks after wildfires throughout the boreal forests. Global Biogeochemical Cycles 34, e2020GB006612.

43. Wardle, D.A., Hornberg, G., Zackrisson, O., Kalela-Brundin, M., and Coomes, D.A. (2003). Long-term effects of wildfire on ecosystem properties across an island area gradient. Science 300, 972–975.

44. Juselius-Rajamäki, T., Väliranta, M., and Korhola, A. (2023). The ongoing lateral expansion of peatlands in Finland. Global Change Biology 29, 7173–7191.

45. Brienen, R.J.W., Phillips, O.L., Feldpausch, T.R., Gloor, E., Baker, T.R., Lloyd, J., Lopez-Gonzalez, G., Monteagudo-Mendoza, A., Malhi, Y., Lewis, S.L., et al. (2015). Long-term decline of the Amazon carbon sink. Nature *519*, 344–348.

46. Scheffer, M., and Carpenter, S.R. (2003). Catastrophic regime shifts in ecosystems: linking theory to observation. Trends in Ecology & Evolution *18*, 648–656.

47. Hanski, I., Turchin, P., Korpimaki, E., and Henttonen, H. (1993). Population oscillations of boreal rodents: regulation by mustelid predators leads to chaos. Nature *364*, 232–235.

48. Geiser, F. (2004). Metabolic rate and body temperature reduction during hibernation and daily torpor. Annual Review of Physiology *66*, 239–274.

49. Eliot, T.S. (1942). Little Gidding (Faber and Faber).

50. Holdridge, L.R. (1967). Life zone ecology (Tropical Science Center).

51. Ryan, M.G. (1991). Effects of climate change on plant respiration. Ecological Applications *1*, 157–167.

52. Mo, L., Crowther, T.W., Maynard, D.S., van den Hoogen, J., Ma, H., Bialic-Murphy, L., Liang, J., de-Miguel, S., Nabuurs, G.-J., and Reich, P.B. (2024). The global distribution and drivers of wood density and their impact on forest carbon stocks. Nature Ecology & Evolution, 8, 2195–2212.

53. White, C.R., and Seymour, R.S. (2003). Mammalian basal metabolic rate is proportional to body mass2/3. Proceedings of the National Academy of Sciences *100*, 4046–4049.

54. Mori, S., Yamaji, K., Ishida, A., Prokushkin, S.G., Masyagina, O.V., Hagihara, A., Rafiqul Hoque, A.T.M., Suwa, A., Osawa, A., Nishizono, T., et al. (2010). Mixed-power scaling of whole-plant respiration from seedlings to giant trees. Proceedings of the National Academy of Sciences *107*, 1447–1451.

55. Chave, J., Coomes, D., Jansen, S., Lewis, S.L., Swenson, N.G., and Zanne, A.E. (2009). Towards a worldwide wood economics spectrum. Ecology Letters *12*, 351–366.

56. Lavers, G. (1983). The strength properties of timber, third edition (Her Majesty's Stationary Office).

57. Rose, M.R. (1994). Evolutionary biology of aging (Oxford University Press).

58. Howes, B., González-Suárez, M., Banks-Leite, C., Bellotto-Trigo, F.C., and Betts, M.G. (2024). A global latitudinal gradient in the proportion of terrestrial vertebrate forest species. Global Ecology and Biogeography *33*, e13854.

Chapter 2

1. Edelman, S.M., and Richards, J.H. (2019). Review of vegetative branching in the palms (Arecaceae). Botanical Review *85*, 40–77.

2. Spector, T., and Putz, F.E. (2006). Biomechanical plasticity facilitates invasion of maritime forests in the southern USA by Brazilian pepper (*Schinus terebinthifolius*). Biological Invasions *8*, 255–260.

3. Scheffer, M., Vergnon, R., Cornelissen, J.H.C., Hantson, S., Holmgren, M., van Nes, E.H., and Xu, C. (2014). Why trees and shrubs but rarely trubs? Trends in Ecology and Evolution *29*, 433–434.

4. Larjavaara, M. (2015). Trees and shrubs differ biomechanically. Trends in Ecology & Evolution *30*, 499–500.

5. FAO (2023). Guidelines and Specifications FRA 2025.

6. Kikuzawa, K. (1991). A cost-benefit analysis of leaf habit and leaf longevity of trees and their geographical pattern. The American Naturalist *138*, 1250–1263.

7. Coley, P.D. (1983). Herbivory and defensive characteristics of tree species in a lowland tropical forest. Ecological Monographs *53*, 209–234.

8. Givnish, T.J. (2002). Adaptive significance of evergreen vs. deciduous leaves: solving the triple paradox. Silva Fennica *36*, 703–743.

9. Wright, I.J., Dong, N., Maire, V., Prentice, I.C., Westoby, M., Díaz, S., Gallagher, R.V., Jacobs, B.F., Kooyman, R., and Law, E.A. (2017). Global climatic drivers of leaf size. Science *357*, 917–921.

10. Niinemets, Ü., Portsmuth, A., Tena, D., Tobias, M., Matesanz, S., and Valladares, F. (2007). Do we underestimate the importance of leaf size in plant economics? disproportional scaling of support costs within the spectrum of leaf physiognomy. Annals of Botany *100*, 283–303.

11. Law, B.E., Falge, E., Gu, L., Baldocchi, D.D., Bakwin, P., Berbigier, P., Davis, K., Dolman, A.J., Falk, M., and Fuentes, J. (2002). Environmental controls over carbon dioxide and water vapor exchange of terrestrial vegetation. Agricultural and Forest Meteorology *113*, 97–120.

12. Van Pelt, R., Sillett, S.C., Kruse, W.A., Freund, J.A., and Kramer, R.D. (2016). Emergent crowns and light-use complementarity lead to global maximum biomass and leaf area in Sequoia sempervirens forests. Forest Ecology Management *375*, 279–308.

13. Lev-Yadun, S. (2011). Bark. In encyclopedia of life sciences (Wiley).

14. Pausas, J.G. (2015). Bark thickness and fire regime. Functional Ecology *29*, 315–327.

15. Niklas, K.J. (1999). The mechanical role of bark. American Journal of Botany *86*, 465–469.

16. Lev-Yadun, S. (2019). Why is the bark of common temperate *Betula* and *Populus* species white? International Journal of Plant Sciences *180*, 632–642.

17. Chen, X., Zhao, P., Zhao, X., Wang, Q., Ouyang, L., Larjavaara, M., Zhu, L., and Ni, G. (2021). Involvement of stem cortical photosynthesis in hydraulic maintenance of Eucalyptus trees and its effect on leaf gas exchange. Environmental and Experimental Botany *186*, 104451.

18. Piovesan, G., and Biondi, F. (2021). On tree longevity. New Phytologist *231*, 1318–1337.

19. Martinez-Ramos, M., and Alvarez-Buylla, E.R. (1998). How old are tropical rain forest trees? Trends in Plant Science *3*, 400–405.

20. Vogel, S. (2012). The life of a leaf (University of Chicago Press).

21. Ledo, A., Paul, K.I., Burslem, D.F., Ewel, J.J., Barton, C., Battaglia, M., Brooksbank, K., Carter, J., Eid, T.H., and England, J.R. (2018). Tree size and climatic water deficit control root to shoot ratio in individual trees globally. New Phytologist *217*, 8–11.

22. Sariyildiz, T., and Anderson, J. (2005). Variation in the chemical composition of green leaves and leaf litters from three deciduous tree species growing on different soil types. Forest Ecology and Management *210*, 303–319.

23. Ometto, J.P., Gorgens, E.B., de Souza Pereira, F.R., Sato, L., de Assis, M.L.R., Cantinho, R., Longo, M., Jacon, A.D., and Keller, M. (2023). A biomass map of the Brazilian Amazon from multisource remote sensing. Scientific Data *10*, 668.

24. Cunha, H.F.V., Andersen, K.M., Lugli, L.F., Santana, F.D., Aleixo, I.F., Moraes, A.M., Garcia, S., Di Ponzio, R., Mendoza, E.O., and Brum, B. (2022). Direct evidence for phosphorus limitation on Amazon forest productivity. Nature *608*, 558–562.

25. Chao, K.J., Phillips, O.L., Gloor, E., Monteagudo, A., Torres-Lezama, A., and Martínez, R.V. (2008). Growth and wood density predict tree mortality in Amazon forests. Journal of Ecology *96*, 281–292.

26. Yang, Y.Y., and Kim, J.G. (2016). The optimal balance between sexual and asexual reproduction in variable environments: a systematic review. Journal of Ecology and Environment *40*, 1–18.

Chapter 3

1. Yang, Y., Saatchi, S.S., Xu, L., Yu, Y.F., Choi, S., Phillips, N., Kennedy, R., Keller, M., Knyazikhin, Y., and Myneni, R.B. (2018). Post-drought decline of the Amazon carbon sink. Nature Communications *9*, 3172.

2. Cochrane, M.A., and Laurance, W.F. (2002). Fire as a large-scale edge effect in Amazonian forests. Journal of Tropical Ecology *18*, 311–325.

3. Martinezramos, M., Alvarezbuylla, E., Sarukhan, J., and Pinero, D. (1988). Treefall age-determination and gap dynamics in a tropical forest. Journal of Ecology *76*, 700–716.

4. Negron-Juarez, R.I., Jenkins, H.S., Raupp, C.F.M., Riley, W.J., Kueppers, L.M., Marra, D.M., Ribeiro, G., Monteiro, M.T.F., Candido, L.A., Chambers, J.Q., and Higuchi, N. (2017). Windthrow variability in Central Amazonia. Atmosphere *8*, 28.

5. Goni, G., DeMaria, M., Knaff, J., Sampson, C., Ginis, I., Bringas, F., Mavume, A., Lauer, C., Lin, I.-I., and Ali, M. (2009). Applications of satellite-derived ocean measurements to tropical cyclone intensity forecasting. Oceanography *22*, 190–197.

6. Pais, C., Gonzalez-Olabarria, J.R., Elimbi Moudio, P., Garcia-Gonzalo, J., González, M.C., and Shen, Z.-J.M. (2023). Global scale coupling of pyromes and fire regimes. Communications Earth & Environment *4*, 267.

7. McCullough, D.G., Werner, R.A., and Neumann, D. (1998). Fire and insects in northern and boreal forest ecosystems of North America. Annual Review of Entomology *43*, 107–127.

8. Rogers, B.M., Soja, A.J., Goulden, M.L., and Randerson, J.T. (2015). Influence of tree species on continental differences in boreal fires and climate feedbacks. Nature Geoscience *8*, 228–234.

9. Fournier, M., Dlouha, J., Jaouen, G., and Almeras, T. (2013). Integrative biomechanics for tree ecology: beyond wood density and strength. Journal of Experimental Botany *64*, 4793–4815.

10. Poorter, L., and Bongers, F. (2006). Leaf traits are good predictors of plant performance across 53 rain forest species. Ecology *87*, 1733–1743.

11. Muller-Landau, H.C. (2004). Interspecific and inter-site variation in wood specific gravity of tropical trees. Biotropica *36*, 20–32.

12. Salguero-Gómez, R., Jones, O.R., Jongejans, E., Blomberg, S.P., Hodgson, D.J., Mbeau-Ache, C., Zuidema, P.A., De Kroon, H., and Buckley, Y.M. (2016). Fast–slow continuum and reproductive strategies structure plant life-history variation worldwide. Proceedings of the National Academy of Sciences *113*, 230–235.

13. Kambach, S., Condit, R., Aguilar, S., Bruelheide, H., Bunyavejchewin, S., Chang-Yang, C.H., Chen, Y.Y., Chuyong, G., Davies, S.J., Ediriweera, S., et al. (2022). Consistency of demographic trade-offs across 13 (sub)tropical forests. Journal of Ecology *110*, 1485–1496.

14. Hirota, M., Holmgren, M., Van Nes, E.H., and Scheffer, M. (2011). Global resilience of tropical forest and savanna to critical transitions. Science *334*, 232–235.

15. Sharam, G., Sinclair, A.R.E., and Turkington, R. (2006). Establishment of broad-leaved thickets in Serengeti, Tanzania: the influence of fire, browsers, grass competition, and elephants. Biotropica *38*, 599–605.

16. Gao, W., and Larjavaara, M. (2024). Wind disturbance in forests: a bibliometric analysis and systematic review. Forest Ecology and Management *564*, 122001.

17. Dongmo Keumo Jiazet, J.H., Dlouha, J., Fournier, M., Moulia, B., Ningre, F., and Constant, T. (2022). No matter how much space and light are available, radial growth distribution in *Fagus sylvatica L.* trees is under strong biomechanical control. Annals of Forest Science *79*, 44.

18. Gardiner, B. (2021). Wind damage to forests and trees: a review with an emphasis on planted and managed forests. Journal of Forest Research *26*, 248–266.

19. Jackson, T.D., Bittencourt, P., Poffley, J., Anderson, J., Muller-Landau, H.C., Ramos, P.A., Rowland, L., and Coomes, D. (2024). Wind shapes the growth strategies of trees in a tropical forest. Ecology Letters *27*, e14527.

20. Ibanez, T., Bauman, D., Aiba, S.i., Arsouze, T., Bellingham, P.J., Birkinshaw, C., Birnbaum, P., Curran, T.J., DeWalt, S.J., Dwyer, J., et al. (2024). Damage to tropical forests caused by cyclones is driven by wind speed but mediated by topographical exposure and tree characteristics. Global Change Biology *30*, e17317.

21. Sherratt, T.N., and Wilkinson, D.M. (2009). Big questions in ecology and evolution (Oxford University Press).

22. Yang, Y.Y., and Kim, J.G. (2016). The optimal balance between sexual and asexual reproduction in variable environments: a systematic review. Journal of Ecology and Environment *40*, 1–18.

23. Larjavaara, M., and Muller-Landau, H.C. (2010). Rethinking the value of high wood density. Functional Ecology *24*, 701–705.

24. Curran, T.J., Gersbach, L.N., Edwards, W., and Krockenberger, A.K. (2008). Wood density predicts plant damage and vegetative recovery rates caused by cyclone disturbance in tropical rainforest tree species of north Queensland, Australia. Austral Ecology *33*, 442–450.

25. Romero, C., and Bolker, B.M. (2008). Effects of stem anatomical and structural traits on responses to stem damage: an experimental study in the Bolivian Amazon. Canadian Journal of Forest Research-Revue Canadienne De Recherche Forestiere *38*, 611–618.

26. Rose, M.R. (1994). Evolutionary biology of aging (Oxford University Press).

27. Lueders, I., Reuken, J., Luther, I., van der Horst, G., Kotze, A., Tordiffe, A., Sieme, H., Jakop, U., and Müller, K. (2024). Effect of age and body condition score on reproductive organ size and sperm parameters in captive male African lion (*Panthera leo*): suggesting a prime breeding age. Theriogenology Wild *5*, 100093.

Chapter 4

1. Christenhusz, M.J., and Byng, J.W. (2016). The number of known plants species in the world and its annual increase. Phytotaxa *261*, 201–217.

2. Cazzolla Gatti, R., Reich, P.B., Gamarra, J.G., Crowther, T., Hui, C., Morera, A., Bastin, J.-F., De-Miguel, S., Nabuurs, G.-J., and Svenning, J.-C. (2022). The number of tree species on Earth. Proceedings of the National Academy of Sciences *119*, e2115329119.

3. Wang, X.Q., and Ran, J.H. (2014). Evolution and biogeography of gymnosperms. Molecular Phylogenetics and Evolution *75*, 24–40.

4. Mutke, J., and Barthlott, W. (2005). Patterns of vascular plant diversity at continental to global scales. Kongelige Danske Videnskabernes Selskab Biologiske Skrifter *55*, 521–537.

5. Armenise, L., Simeone, M.C., Piredda, R., and Schirone, B. (2012). Validation of DNA barcoding as an efficient tool for taxon identification and detection of species diversity in Italian conifers. European Journal of Forest Research *131*, 1337–1353.

6. Santoro, M., Cartus, O., Carvalhais, N., Rozendaal, D.M., Avitabile, V., Araza, A., De Bruin, S., Herold, M., Quegan, S., Rodríguez-Veiga, P., et al. (2021). The global forest above-ground biomass pool for 2010 estimated from high-resolution satellite observations. Earth System Science Data *13*, 3927–3950.

7. FAO (2024). The state of the world's forests 2024.

8. Eldredge, N., and Gould, S.J. (1972). Punctuated equilibria: an alternative to phyletic gradualism. Models in Paleobiology *82*, 115.

9. Johnson, N.A., Lahti, D.C., and Blumstein, D.T. (2012). Combating the assumption of evolutionary progress: lessons from the decay and loss of traits. Evolution: Education and Outreach *5*, 128–138.

10. Ashton, M.S., and Kelty, M.J. (2018). The practice of silviculture: applied forest ecology (John Wiley & Sons).

11. Niklas, K.J., and Spatz, H.C. (2012). Plant physics (University of Chicago Press).

12. Benton, M.J., Wilf, P., and Sauquet, H. (2022). The Angiosperm Terrestrial Revolution and the origins of modern biodiversity. New Phytologist *233*, 2017–2035.

13. Nicotra, A.B., Leigh, A., Boyce, C.K., Jones, C.S., Niklas, K.J., Royer, D.L., and Tsukaya, H. (2011). The evolution and functional significance of leaf shape in the angiosperms. Functional Plant Biology *38*, 535–552.

14. Jucker, T., Fischer, F., Chave, J., Coomes, D., Caspersen, J., Ali, A., Panzou, G., Feldpausch, T., Falster, D., Usoltsev, V., et al. (2024). The global spectrum of tree crown architecture. Nature Communications 16, 4876.

15. Li, F., Qian, H., Sardans, J., Amishev, D.Y., Wang, Z., Zhang, C., Wu, T., Xu, X., Tao, X., and Huang, X. (2024). Evolutionary history shapes variation of wood density of tree species across the world. Plant Diversity *46*, 283–293.

16. Mo, L., Crowther, T.W., Maynard, D.S., van den Hoogen, J., Ma, H., Bialic-Murphy, L., Liang, J., de-Miguel, S., Nabuurs, G.-J., Reich, P.B., et al. (2024). The global distribution and drivers of wood density and their impact on forest carbon stocks. Nature Ecology & Evolution, 2195–2212.

17. Muller-Landau, H.C. (2004). Interspecific and inter-site variation in wood specific gravity of tropical trees. Biotropica *36*, 20–32.

18. Larjavaara, M., and Muller-Landau, H.C. (2010). Rethinking the value of high wood density. Functional Ecology *24*, 701–705.

19. Coomes, D.A., Allen, R.B., Bentley, W.A., Burrows, L.E., Canham, C.D., Fagan, L., Forsyth, D.M., Gaxiola-alcantar, A., Parfitt, R.L., Ruscoe, W., et al. (2005). The hare, the tortoise and the crocodile: the ecology of angiosperm dominance, conifer persistence and fern filtering. Journal of Ecology *93*, 918–935.

20. Chave, J., Coomes, D., Jansen, S., Lewis, S.L., Swenson, N.G., and Zanne, A.E. (2009). Towards a worldwide wood economics spectrum. Ecology Letters *12*, 351–366.

21. Liu, J., Xia, S., Zeng, D., Liu, C., Li, Y., Yang, W., Yang, B., Zhang, J., Slik, F., and Lindenmayer, D.B. (2022). Age and spatial distribution of the world's oldest trees. Conservation Biology *36*, e13907.

Chapter 5

1. Guariguata, M.R. (1998). Response of forest tree saplings to experimental mechanical damage in lowland Panama. Forest Ecology and Management *102*, 103–111.

2. Muller-Landau, H., and Pacala, S.W. (2020). What determines the abundance of lianas and vines? In Unsolved problems in ecology (Princeton University Press).

3. Ingwell, L.L., Wright, S.J., Becklund, K.K., Hubbell, S.P., and Schnitzer, S.A. (2010). The impact of lianas on 10 years of tree growth and mortality on Barro Colorado Island, Panama. Journal of Ecology *98*, 879–887.

4. Larjavaara, M., Auvinen, M., Kantola, A., and Mäkelä, A. (2021). Wind and gravity in shaping Picea trunks. Trees Structure & Functioning *35*, 1587–1599.

5. Pruyn, M.L., Gartner, B.L., and Harmon, M.E. (2002). Respiratory potential in sapwood of old versus young ponderosa pine trees in the Pacific Northwest. Tree Physiology *22*, 105–116.

6. Pausas, J.G. (2017). Bark thickness and fire regime: another twist. New Phytologist *213*, 13–15.

7. Lindmark, M., Sunnerheim, K., and Jonsson, B.G. (2020). Natural browsing repellent to protect Scots pine *Pinus sylvestris* from European moose *Alces alces*. Forest Ecology and Management *474*, 118347.

8. Harley, M.M. (2006). A summary of fossil records for Arecaceae. Botanical Journal of the Linnean Society *151*, 39–67.

9. Biswas, C., and Johri, B.M. (2013). The gymnosperms (Springer Science & Business Media).

10. Farjon, A. (2010). A handbook of the world's conifers (Brill).

11. Bouchenak-Khelladi, Y., Verboom, G.A., Savolainen, V., and Hodkinson, T.R. (2010). Biogeography of the grasses (Poaceae): a phylogenetic approach to reveal evolutionary history in geographical space and geological time. Botanical Journal of the Linnean Society *162*, 543–557.

12. Wei, Q., Guo, L., Jiao, C., Fei, Z.J., Chen, M., Cao, J.J., Ding, Y.L., and Yuan, Q.S. (2019). Characterization of the developmental dynamics of the elongation of a bamboo internode during the fast growth stage. Tree Physiology *39*, 1201–1214.

13. Larjavaara, M. (2015). Trees and shrubs differ biomechanically. Trends in Ecology & Evolution *30*, 499–500.

14. Jucker, T., Fischer, F., Chave, J., Coomes, D., Caspersen, J., Ali, A., Panzou, G., Feldpausch, T., Falster, D., Usoltsev, V., et al. (2024). The global spectrum of tree crown architecture. Nature Communications 16, 4876.

15. Jucker, T., Caspersen, J., Chave, J., Antin, C., Barbier, N., Bongers, F., Dalponte, M., van Ewijk, K.Y., Forrester, D.I., Haeni, M., et al. (2017). Allometric equations for integrating remote sensing imagery into forest monitoring programmes. Global Change Biology *23*, 177–190.

16. Ennos, A.R. (1993). The scaling of root anchorage. Journal of Theoretical Biology *161*, 61–75.

17. Paul, K.I., Larmour, J., Specht, A., Zerihun, A., Ritson, P., Roxburgh, S.H., Sochacki, S., Lewis, T., Barton, C.V.M., England, J.R., et al. (2019). Testing the generality of below-ground biomass allometry across plant functional types. Forest Ecology and Management *432*, 102–114.

18. O'Brien, E.E., Brown, J.S., and Moll, J.D. (2007). Roots in space: a spatially explicit model for below-ground competition in plants. Philosophical Transactions of the Royal Society B. *274*, 929–934.

19. Peltola, H., Kellomaki, S., Hassinen, A., and Granander, M. (2000). Mechanical stability of Scots pine, Norway spruce and birch: an analysis of tree-pulling experiments in Finland. Forest Ecology and Management *135*, 143–153.

20. Niklas, K.J. (1992). Plant biomechanics: an engineering approach to plant form and function (University of Chicago Press).

21. Larjavaara, M., and Muller-Landau, H.C. (2012). Still rethinking the value of high wood density. American Journal of Botany *99*, 165–168.

22. Larjavaara, M. (2021). What would a tree say about its size? Frontiers in Ecology and Evolution *8*, 564302.

23. Muller-Landau, H.C. (2004). Interspecific and inter-site variation in wood specific gravity of tropical trees. Biotropica *36*, 20–32.

24. Hietz, P., Valencia, R., and Joseph Wright, S. (2013). Strong radial variation in wood density follows a uniform pattern in two neotropical rain forests. Functional Ecology *27*, 684–692.

25. Richter, W. (1984). A structural approach to the function of buttresses of quararibea-asterolepis. Ecology *65*, 1429–1435.

26. Rader, A.M., Cottrell, A., Kudla, A., Lum, T., Henderson, D., Karandikar, H., and Letcher, S.G. (2020). Tree functional traits as predictors of microburst-associated treefalls in tropical wet forests. Biotropica *52*, 410–414.

27. Gardiner, B., Barnett, J., Saranpää, P., and Gril, J. (2014). The biology of reaction wood (Springer).

28. Telewski, F.W. (2012). Is windswept tree growth negative thigmotropism? Plant Science *184*, 20–28.

29. Thurner, M., Beer, C., Crowther, T., Falster, D., Manzoni, S., Prokushkin, A., and Schulze, E.D. (2019). Sapwood biomass carbon in northern boreal and temperate forests. Global Ecology and Biogeography *28*, 640–660.

30. Putz, F.E. (1983). Liana biomass and leaf area of a "tierra firme" forest in the Rio Negro Basin, Venezuela. Biotropica *15*, 185–189.

31. Morikawa, T., Ashitani, T., Sekine, N., Kusumoto, N., and Takahashi, K. (2012). Bioactivities of extracts from *Chamaecyparis obtusa* branch heartwood. Journal of Wood Science *58*, 544–549.

32. Larjavaara, M., and Muller-Landau, H.C. (2010). Rethinking the value of high wood density. Functional Ecology *24*, 701–705.

33. Gleason, S.M., Westoby, M., Jansen, S., Choat, B., Hacke, U.G., Pratt, R.B., Bhaskar, R., Brodribb, T.J., Bucci, S.J., Cao, K.F., et al. (2016). Weak tradeoff between xylem safety and xylem-specific hydraulic efficiency across the world's woody plant species. New Phytologist *209*, 123–136.

34. Falster, D.S., and Westoby, M. (2003). Plant height and evolutionary games. Trends in Ecology & Evolution *18*, 337–343.

35. Kuuluvainen, T., and Pukkala, T. (1987). Effect of crown shape and tree distribution on the spatial distribution of shade. Agricultural and Forest Meteorology *40*, 215–231.

36. Horn, H.S. (1971). The adaptive geometry of trees (Princeton University Press).

Chapter 6

1. Wang, Z.H., Li, Y.Q., Su, X.Y., Tao, S.L., Feng, X., Wang, Q.G., Xu, X.T., Liu, Y.P., Michaletz, S.T., Shrestha, N., et al. (2019). Patterns and ecological determinants of woody plant height in eastern Eurasia and its relation to primary productivity. Journal of Plant Ecology *12*, 791–803.

2. Slik, J.W.F., Paoli, G., McGuire, K., Amaral, I., Barroso, J., Bastian, M., Blanc, L., Bongers, F., Boundja, P., Clark, C., et al. (2013). Large trees drive forest aboveground biomass variation in moist lowland forests across the tropics. Global Ecology and Biogeography 22, 1261–1271.

3. Kenzo, T., Inoue, Y., Yoshimura, M., Yamashita, M., Tanaka-Oda, A., and Ichie, T. (2015). Height-related changes in leaf photosynthetic traits in diverse Bornean tropical rain forest trees. Oecologia 177, 191–202.

4. Wright, S.J., Turner, B.L., Yavitt, J.B., Harms, K.E., Kaspari, M., Tanner, E.V., Bujan, J., Griffin, E.A., Mayor, J.R., Pasquini, S.C., et al. (2018). Plant responses to fertilization experiments in lowland, species-rich, tropical forests. Ecology 99, 1129–1138.

5. Schulte-Uebbing, L., and de Vries, W. (2018). Global-scale impacts of nitrogen deposition on tree carbon sequestration in tropical, temperate, and boreal forests: a meta-analysis. Global Change Biology 24, e416–e431.

6. Mäkelä, A., and Valentine, H.T. (2020). Models of tree and stand dynamics (Springer).

7. Larjavaara, M., Chen, X., and Luo, M. (2024). A temperature-based model of biomass accumulation in humid forests of the world. Frontiers in Forests and Global Change 7, 1142209.

8. Mori, S., Yamaji, K., Ishida, A., Prokushkin, S.G., Masyagina, O.V., Hagihara, A., Rafiqul Hoque, A.T.M., Suwa, A., Osawa, A., Nishizono, T., et al. (2010). Mixed-power scaling of whole-plant respiration from seedlings to giant trees. Proceedings of the National Academy of Sciences 107, 1447–1451.

9. Ryan, M.G., Phillips, N., and Bond, B.J. (2006). The hydraulic limitation hypothesis revisited. Plant Cell and Environment 29, 367–381.

10. Anderson-Teixeira, K.J., Herrmann, V., Morgan, R.B., Bond-Lamberty, B., Cook-Patton, S.C., Ferson, A.E., Muller-Landau, H., and Wang, M. (2021). Carbon cycling in mature and regrowth forests globally. Environmental Research Letters 16, 053009.

11. Koenig, W.D., and Knops, J.M.H. (1998). Scale of mast-seeding and tree-ring growth. Nature 396, 225–226.

12. Yoda, K., Kira, T., Ogawa, H., and Hozumi, K. (1963). Self-thinning in overcrowded pure stands under cultivated and natural conditions. Journal of Biology, Osaka City University 14, 107–129.

13. Larjavaara, M. (2010). Maintenance cost, toppling risk and size of trees in a self-thinning stand. Journal of Theoretical Biology 265, 63–67.

14. Bader, M.K.F., Leuzinger, S., Keel, S.G., Siegwolf, R.T., Hagedorn, F., Schleppi, P., and Körner, C. (2013). Central European hardwood trees in a high-CO_2 future: synthesis of an 8-year forest canopy CO_2 enrichment project. Journal of Ecology 101, 1509–1519.

15. Norby, R.J., and Zak, D.R. (2011). Ecological lessons from free-air CO_2 enrichment (FACE) experiments. Annual Review of Ecology, Evolution, and Systematics 42, 181–203.

16. De La Vega, E., Chalk, T.B., Wilson, P.A., Bysani, R.P., and Foster, G.L. (2020). Atmospheric CO2 during the Mid-Piacenzian Warm Period and the M2 glaciation. Scientific Reports *10*, 1–8.

17. Sala, A., Fouts, W., and Hoch, G. (2011). Carbon storage in trees: does relative carbon supply decrease with tree size? In Size-and age-related changes in tree structure and function (Springer).

18. Cheesman, A.W., and Winter, K. (2013). Elevated night-time temperatures increase growth in seedlings of two tropical pioneer tree species. New Phytologist *197*, 1185–1192.

19. Jucker, T., Fischer, F., Chave, J., Coomes, D., Caspersen, J., Ali, A., Panzou, G., Feldpausch, T., Falster, D., Usoltsev, V., et al. (2024). The global spectrum of tree crown architecture. Nature Communications *16*, 4876

20. Chave, J., Coomes, D., Jansen, S., Lewis, S.L., Swenson, N.G., and Zanne, A.E. (2009). Towards a worldwide wood economics spectrum. Ecology Letters *12*, 351–366.

21. Larjavaara, M. (2021). What would a tree say about its size? Frontiers in Ecology and Evolution *8*, 564302.

22. Wohlleben, P. (2016). The hidden life of trees: what they feel, how they communicate—Discoveries from a secret world (Greystone Books).

23. Detto, M., Visser, M.D., Wright, S.J., and Pacala, S.W. (2019). Bias in the detection of negative density dependence in plant communities. Ecology Letters *22*, 1923–1939.

24. File, A.L., Murphy, G.P., and Dudley, S.A. (2012). Fitness consequences of plants growing with siblings: reconciling kin selection, niche partitioning and competitive ability. Philosophical Transactions of the Royal Society B *279*, 209–218.

Chapter 7

1. Chen, X., Luo, M., Kang, Y., Zhao, P., Tang, Z., Meng, Y., Huang, L., Guo, Y., Lu, X., Ouyang, L., et al. (2023). Comparison between the stem and leaf photosynthetic productivity in *Eucalyptus urophylla* plantations with different age. Planta *257*, 56.

2. Dalling, J.W., Flores III, M.R., and Heineman, K.D. (2024). Wood nutrients: Underexplored traits with functional and biogeochemical consequences. New Phytologist *244*, 1694–1708.

3. Chapotin, S.M., Razanameharizaka, J.H., and Holbrook, N.M. (2006). Baobab trees (*Adansonia*) in Madagascar use stored water to flush new leaves but not to support stomatal opening before the rainy season. New Phytologist *169*, 549–559.

4. Thompson, M.V., and Holbrook, N.M. (2003). Application of a single-solute non-steady-state phloem model to the study of long-distance assimilate transport. Journal of Theoretical Biology *220*, 419–455.

5. Koch, G.W., Sillett, S.C., Jennings, G.M., and Davis, S.D. (2004). The limits to tree height. Nature *428*, 851–854.

6. Givnish, T.J., Wong, S.C., Stuart-Williams, H., Holloway-Phillips, M., and Farquhar, G.D. (2014). Determinants of maximum tree height in Eucalyptus species along a rainfall gradient in Victoria, Australia. Ecology *95*, 2991–3007.

7. Tng, D.Y.P., Williamson, G.J., Jordan, G.J., and Bowman, D.M.J.S. (2012). Giant eucalypts—globally unique fire-adapted rain-forest trees? New Phytologist *196*, 1001–1014.

8. Grossiord, C., Buckley, T.N., Cernusak, L.A., Novick, K.A., Poulter, B., Siegwolf, R.T., Sperry, J.S., and McDowell, N.G. (2020). Plant responses to rising vapor pressure deficit. New Phytologist *226*, 1550–1566.

9. Mitchell, S. (2013). Wind as a natural disturbance agent in forests: a synthesis. Forestry *86*, 147–157.

10. Larjavaara, M. (2015). Trees and shrubs differ biomechanically. Trends in Ecology & Evolution *30*, 499–500.

11. Gardiner, B. (2021). Wind damage to forests and trees: a review with an emphasis on planted and managed forests. Journal of Forest Research *26*, 248–266.

12. Larjavaara, M., Auvinen, M., Kantola, A., and Mäkelä, A. (2021). Wind and gravity in shaping *Picea* trunks. Trees Structure & Functioning. 10.1007/s00468-021-02138-3.

13. West, P.W. (2014). Growing plantation forests (Springer).

14. Shinozaki, K., Yoda, K., Hozumi, K., and Kira, T. (1964). A quantitative analysis on plant form: The pipe model theory. I Basic analyses. Japanese Journal of Ecology *14*, 97–105.

15. Zhang, W.-P., Zhao, L., Larjavaara, M., Morris, E.C., Sterck, F.J., and Wang, G.-X. (2020). Height-diameter allometric relationships for seedlings and trees across China. Acta Oecologica *108*, 103621.

16. Larjavaara, M. (2010). Maintenance cost, toppling risk and size of trees in a self-thinning stand. Journal of Theoretical Biology *265*, 63–67.

17. Niklas, K.J., and Spatz, H.C. (2012). Plant Physics (University of Chicago Press).

18. McMahon, T. (1973). Size and shape in biology. Science *179*, 1201–1204.

19. Enquist, B.J., and Niklas, K.J. (2001). Invariant scaling relations across tree-dominated communities. Nature *410*, 655–660.

20. Sprengel, C. (1837). Die Bodenkunde oder die Lehre vom Boden: nebst einer vollständigen Anleitung zur chemischen Analyse der Ackererden: ein Handbuch für Landwirthe, Forstmänner (Müller).

21. Morris, D. (2009). The human zoo (Random House).

22. Shook, R.P., Hand, G.A., Paluch, A.E., Wang, X., Moran, R., Hébert, J.R., Lavie, C.J., and Blair, S.N. (2014). Moderate cardiorespiratory fitness is positively associated with resting metabolic rate in young adults. Mayo Clinic Proceedings *89*, 763–771.

23. Dlouhá, J., Moulia, B., Fournier, M., Badel, E., and Constant, T. (2025). Beyond the perception of wind only as a meteorological hazard: importance of mechanobiology for biomass allocation, forest ecology and management. Annals of Forest Science *82*, 1.

Chapter 8

1. Sillett, S.C., Van Pelt, R., Koch, G.W., Ambrose, A.R., Carroll, A.L., Antoine, M.E., and Mifsud, B.M. (2010). Increasing wood production through old age in tall trees. Forest Ecology and Management *259*, 976–994.

2. Cheng, Z., Aakala, T., and Larjavaara, M. (2023). Elevation, aspect, and slope influence woody vegetation structure and composition but not species richness in a human-influenced landscape in northwestern Yunnan, China. Frontiers in Forests and Global Change 6, 1187724.

3. Diamond, J.M. (1998). Guns, germs, and steel (Vintage Publishing).

4. Dawson, T.E. (1998). Fog in the California redwood forest: ecosystem inputs and use by plants. Oecologia *117*, 476–485.

5. Limm, E.B., Simonin, K.A., Bothman, A.G., and Dawson, T.E. (2009). Foliar water uptake: a common water acquisition strategy for plants of the redwood forest. Oecologia *161*, 449–459.

6. Gora, E.M., Burchfield, J.C., Muller-Landau, H.C., Bitzer, P.M., and Yanoviak, S.P. (2020). Pantropical geography of lightning-caused disturbance and its implications for tropical forests. Global Change Biology 26, 5017–5026.

7. Kaplan, J.O., and Lau, K.H.-K. (2022). World wide lightning location network (WWLLN) global lightning climatology (WGLC) and time series, 2022 update. Earth System Science Data 14, 5665–5670.

8. Shenkin, A., Chandler, C., Boyd, D., Jackson, T., bin Jami, J., Disney, M., Majalap, N., Nilus, R., Foody, G., Reynolds, et al. (2019). The world's tallest tropical tree in three dimensions. Frontiers in Forests and Global Change 2, 32.

9. Ren, Y., Li, C., Chau, K., Fan, G., Xu, G., Yang, H., Cheng, K., Hua, F., Hu, R., and Shi, X. (2024). Conserving the primary forests in the Yarlung Tsangpo Grand Canyon for people and nature. Nature Ecology & Evolution 8, 837–839.

10. Slik, J.W.F., Paoli, G., McGuire, K., Amaral, I., Barroso, J., Bastian, M., Blanc, L., Bongers, F., Boundja, P., Clark, C., et al. (2013). Large trees drive forest aboveground biomass variation in moist lowland forests across the tropics. Global Ecology and Biogeography *22*, 1261–1271.

11. Keeling, H.C., and Phillips, O.L. (2007). The global relationship between forest productivity and biomass. Global Ecology and Biogeography *16*, 618–631.

12. Tao, S., Guo, Q., Li, C., Wang, Z., and Fang, J. (2016). Global patterns and determinants of forest canopy height. Ecology *97*, 3265–3270.

13. Larjavaara, M., and Muller-Landau, H.C. (2013). Corrigendum on: Temperature explains global variation in biomass among humid old-growth forests. Global Ecology and Biogeography *22*, 772.

14. Larjavaara, M., Chen, X., and Luo, M. (2024). A temperature-based model of biomass accumulation in humid forests of the world. Frontiers in Forests and Global Change, *7*, 1142209.

15. Larjavaara, M. (2014). The world's tallest trees grow in thermally similar climates. New Phytologist *202*, 344–349.

16. Fine, P.V., and Ree, R.H. (2006). Evidence for a time-integrated species-area effect on the latitudinal gradient in tree diversity. The American Naturalist *168*, 796–804.

17. Hemp, A., Zimmermann, R., Remmele, S., Pommer, U., Berauer, B., Hemp, C., and Fischer, M. (2017). Africa's highest mountain harbours Africa's tallest trees. Biodiversity and Conservation *26*, 103–113.

18. Girardin, M.P., Bouriaud, O., Hogg, E.H., Kurz, W., Zimmermann, N.E., Metsaranta, J.M., de Jong, R., Frank, D.C., Esper, J., Buntgen, U., et al. (2016). No growth stimulation of Canada's boreal forest under half-century of combined warming and CO_2 fertilization. Proceedings of the National Academy of Sciences *113*, E8406–E8414.

19. Sillett, S.C., Kramer, R.D., Van Pelt, R., Carroll, A.L., Campbell-Spickler, J., and Antoine, M.E. (2021). Comparative development of the four tallest conifer species. Forest Ecology and Management *480*, 118688.

20. Akaike, H. (1998). Information theory and an extension of the maximum likelihood principle. In Selected papers of Hirotugu Akaike (Springer).

21. Mäkelä, A., and Valentine, H.T. (2020). Models of tree and stand dynamics (Springer).

22. Huntzinger, D., Michalak, A., Schwalm, C., Ciais, P., King, A., Fang, Y., Schaefer, K., Wei, Y., Cook, R., Fisher, J., et al. (2017). Uncertainty in the response of terrestrial carbon sink to environmental drivers undermines carbon-climate feedback predictions. Scientific Reports *7*, 4765.

23. Bar-On, Y.M., Li, X., O'sullivan, M., Wigneron, J.-P., Sitch, S., Ciais, P., Frankenberg, C., and Fischer, W.W. (2025). Recent gains in global terrestrial carbon stocks are mostly stored in nonliving pools. Science *387*, 1291–1295.

24. Purves, D., and Pacala, S. (2008). Predictive models of forest dynamics. Science *320*, 1452–1453.

Chapter 9

1. Moris, J.V., Álvarez-Álvarez, P., Conedera, M., Dorph, A., Hessilt, T.D., Hunt, H.G., Libonati, R., Menezes, L.S., Müller, M.M., and Pérez-Invernón, F.J. (2023). A global database on holdover time of lightning-ignited wildfires. Earth System Science Data *15*, 1151–1163.

2. Sandom, C., Faurby, S., Sandel, B., and Svenning, J.C. (2014). Global late Quaternary megafauna extinctions linked to humans, not climate change. Philosophical Transactions of the Royal Society B. *281*, 20133254.

3. Barlow, C. (2008). The ghosts of evolution: nonsensical fruit, missing partners, and other ecological anachronisms (Basic Books).

4. Peres, C.A., Emilio, T., Schietti, J., Desmouliere, S.J.M., and Levi, T. (2016). Dispersal limitation induces long-term biomass collapse in overhunted Amazonian forests. Proceedings of the National Academy of Sciences *113*, 892–897.

5. Larjavaara, M. (2021). What would a tree say about its size? Frontiers in Ecology and Evolution *8*, 564302.

6. Mansson, J., Kalen, C., Kjellander, P., Andren, H., and Smith, H. (2007). Quantitative estimates of tree species selectivity by moose (*Alces alces*) in a forest landscape. Scandinavian Journal of Forest Research 22, 407–414.

7. Roebroeks, W., and Villa, P. (2011). On the earliest evidence for habitual use of fire in Europe. Proceedings of the National Academy of Sciences 108, 5209–5214.

8. Zeder, M.A. (2012). The domestication of animals. Journal of Anthropological Research 68, 161–190.

9. Wood, J.W. (2017). Dynamics of human reproduction: biology, biometry, demography (Routledge).

10. Malthus, T.R. (1798). An essay on the principle of population (J. Johnson).

11. Ramankutty, N., Mehrabi, Z., Waha, K., Jarvis, L., Kremen, C., Herrero, M., and Rieseberg, L.H. (2018). Trends in global agricultural land use: implications for environmental health and food security. Annual Review of Plant Biology 69, 789–815.

12. Mather, A.S., and Needle, C. (1998). The forest transition: a theoretical basis. Area 30, 117–124.

13. Jayathilake, H.M., Prescott, G.W., Carrasco, L.R., Rao, M., and Symes, W.S. (2021). Drivers of deforestation and degradation for 28 tropical conservation landscapes. Ambio 50, 215–228.

14. Kummer, D.M. (1992). Deforestation in the postwar Philippines (University of Chicago Press).

15. Verissimo, A., Barreto, P., Tarifa, R., and Uhl, C. (1995). Extraction of a high-value natural resource in Amazonia: the case of mahogany. Forest Ecology and Management 72, 39–60.

16. West, C., Rabeschini, G., Singh, C., Kastner, T., Bastos Lima, M., Dermawan, A., Croft, S., and Persson, U.M. (2025). The global deforestation footprint of agriculture and forestry. Nature Reviews Earth & Environment 6, 325–341.

17. Diamond, J. (2005). Collapse: how societies choose to fail or survive (Penguin).

18. Allsen, T.T. (2011). The royal hunt in Eurasian history (University of Pennsylvania Press).

19. Tasanen, T. (2004). Läksi puut ylenemähän: metsien hoidon historia Suomessa keskiajalta metsäteollisuuden läpimurtoon 1870-luvulla (Metsäntutkimuslaitoksen tiedonantoja).

20. Cheng, K., Yang, H., Tao, S., Su, Y., Guan, H., Ren, Y., Hu, T., Li, W., Xu, G., and Chen, M. (2024). Carbon storage through China's planted forest expansion. Nature Communications 15, 4106.

21. Meyfroidt, P., and Lambin, E.F. (2008). Forest transition in Vietnam and its environmental impacts. Global Change Biology 14, 1319–1336.

22. Larjavaara, M. (2012). Democratic less-developed countries cause global deforestation. International Forestry Review 14, 299–313.

23. Hardin, G. (1968). The tragedy of the commons: the population problem has no technical solution; it requires a fundamental extension in morality. Science 162, 1243–1248.

24. Domínguez-Rodrigo, M. (2014). Is the "savanna hypothesis" a dead concept for explaining the emergence of the earliest hominins? Current Anthropology 55, 59–81.

25. Ellis, E.C. (2021). Land use and ecological change: a 12,000-year history. Annual Review of Environment and Resources 46, 1–33.

26. Zong, Y., Chen, Z., Innes, J.B., Chen, C., Wang, Z., and Wang, H. (2007). Fire and flood management of coastal swamp enabled first rice paddy cultivation in east China. Nature 449, 459–462.

27. Iriarte, J., Elliott, S., Maezumi, S.Y., Alves, D., Gonda, R., Robinson, M., de Souza, J.G., Watling, J., and Handley, J. (2020). The origins of Amazonian landscapes: plant cultivation, domestication and the spread of food production in tropical South America. Quaternary Science Reviews 248, 106582.

28. Mello, P.A. (2017). Democratic peace theory. In The SAGE encyclopedia of war: Social science perspectives.

Chapter 10

1. Nghiem, N. (2014). Optimal rotation age for carbon sequestration and biodiversity conservation in Vietnam. Forest Policy and Economics 38, 56–64.

2. Evans, J., and Turnbull, J.W. (2004). Plantation forestry in the tropics: the role, silviculture, and use of planted forests for industrial, social, environmental, and agroforestry purposes, 3rd ed. (Oxford University Press).

3. Mäkelä, A., and Valentine, H.T. (2020). Models of tree and stand dynamics (Springer).

4. Larjavaara, M., Chen, X., and Luo, M. (2024). A temperature-based model of biomass accumulation in humid forests of the world. Frontiers in Forests and Global Change 7, 1142209.

5. Hall, J.S., and Ashton, M. (2016). Guide to early growth and survival in plantations of 64 tree species native to Panama and the neotropics (Smithsonian Tropical Research Institute).

6. Valverde-Barrantes, O.J., Smemo, K.A., Feinstein, L.M., Kershner, M.W., and Blackwood, C.B. (2013). The distribution of below-ground traits is explained by intrinsic species differences and intraspecific plasticity in response to root neighbours. Journal of Ecology 101, 933–942.

7. Narwal, S., Hoagland, R.E., Dilday, R., and Roger, M.R. (2012). Allelopathy in ecological agriculture and forestry. In Allelopathy in ecological agriculture and forestry (Springer).

8. Larjavaara, M. (2021). What would a tree say about its size? Frontiers in Ecology and Evolution 8, 564302.

9. Crous, C.J., Burgess, T.I., Le Roux, J.J., Richardson, D.M., Slippers, B., and Wingfield, M.J. (2017). Ecological disequilibrium drives insect pest and pathogen accumulation in non-native trees. Aob Plants 9, plw081.

10. Stephens, S.S., and Wagner, M.R. (2007). Forest plantations and biodiversity: a fresh perspective. Journal of Forestry 105, 307–313.

11. Jucker, T., Fischer, F., Chave, J., Coomes, D., Caspersen, J., Ali, A., Panzou, G., Feldpausch, T., Falster, D., and Usoltsev, V. (2024). The global spectrum of tree crown architecture. Nature Communications *16*, 4876.

12. FAO (2024). The state of the world's forests 2024.

13. Valkonen, S., and Valsta, L. (2001). Productivity and economics of mixed two-storied spruce and birch stands in Southern Finland simulated with empirical models. Forest Ecology and Management *140*, 133–149.

14. Larjavaara, M. (2008). A review on benefits and disadvantages of tree diversity. The Open Forest Science Journal *1*, 24–26.

15. Sharman, P., and Wilson, A.J. (2023). Genetic improvement of speed across distance categories in thoroughbred racehorses in Great Britain. Heredity *131*, 79–85.

16. Rüger, N., Condit, R., Dent, D.H., DeWalt, S.J., Hubbell, S.P., Lichstein, J.W., Lopez, O.R., Wirth, C., and Farrior, C.E. (2020). Demographic trade-offs predict tropical forest dynamics. Science *368*, 165–168.

17. Hua, F., Bruijnzeel, L.A., Meli, P., Martin, P.A., Zhang, J., Nakagawa, S., Miao, X., Wang, W., McEvoy, C., Peña-Arancibia, et al. (2022). The biodiversity and ecosystem service contributions and trade-offs of forest restoration approaches. Science *376*, 839–844.

18. Hua, F., Wang, X., Zheng, X., Fisher, B., Wang, L., Zhu, J., Tang, Y., Yu, D.W., and Wilcove, D.S. (2016). Opportunities for biodiversity gains under the world's largest reforestation programme. Nature Communications *7*, 12717.

19. Putz, F.E., and Romero, C. (2015). Futures of tropical production forests (CIFOR).

Chapter 11

1. Peng, L., Searchinger, T.D., Zionts, J., and Waite, R. (2023). The carbon costs of global wood harvests. Nature *620*, 110–115.

2. Shukla, P., Skea, J., Calvo Buendia, E., Masson-Delmotte, V., Pörtner, H., Roberts, D., Zhai, P., Slade, R., Connors, S., Van Diemen, R., et al. (2019). Climate change and land: an IPCC special report on climate change, desertification, land degradation, sustainable land management, food security, and greenhouse gas fluxes in terrestrial ecosystems (IPCC).

3. Pan, Y., Birdsey, R.A., Phillips, O.L., Houghton, R.A., Fang, J., Kauppi, P.E., Keith, H., Kurz, W.A., Ito, A., Lewis, S.L., et al. (2024). The enduring world forest carbon sink. Nature *631*, 563–569.

4. Martin, A.R., Doraisami, M., and Thomas, S.C. (2018). Global patterns in wood carbon concentration across the world's trees and forests. Nature Geoscience *11*, 915–920.

5. Lorenz, K., and Lal, R. (2005). The depth distribution of soil organic carbon in relation to land use and management and the potential of carbon sequestration in subsoil horizons. Advances in Agronomy 88, 35–66.

6. Pan, Y., Birdsey, R.A., Fang, J., Houghton, R., Kauppi, P.E., Kurz, W.A., Phillips, O.L., Shvidenko, A., Lewis, S.L., Canadell, J.G., et al. (2011). A large and persistent carbon sink in the world's forests. Science *333*, 988–993.

7. Gora, E.M., Kneale, R.C., Larjavaara, M., and Muller-Landau, H.C. (2019). Dead wood necromass in a moist tropical forest: stocks, fluxes, and spatiotemporal variability. Ecosystems 22, 1189–1205.

8. Larjavaara, M., Kanninen, M., Gordillo, H., Koskinen, J., Kukkonen, M., Käyhkö, N., Larson, A.M., and Wunder, S. (2018). Global variation in the cost of increasing ecosystem carbon. Nature Climate Change 8, 38.

9. Pukkala, T. (2018). Carbon forestry is surprising. Forest Ecosystems 5, 11.

10. Hurmekoski, E., Smyth, C.E., Stern, T., Verkerk, P.J., and Asada, R. (2021). Substitution impacts of wood use at the market level: a systematic review. Environmental Research Letters 16, 123004.

11. Putz, F.E., Cayetano, D.T., Belair, E.P., Ellis, P.W., Roopsind, A., Griscom, B.W., Finlayson, C., Finkral, A., Cho, P.P., and Romero, C. (2023). Liana cutting in selectively logged forests increases both carbon sequestration and timber yields. Forest Ecology and Management 539, 121038.

12. Larjavaara, M., Mehtätalo, L., Lehtonen, A., and Räty, M. (2023). Comment on "Releasing global forests from human management: How much more carbon could be stored?". Science eLetter.

13. Assmuth, A., and Tahvonen, O. (2018). Optimal carbon storage in even- and uneven-aged forestry. Forest Policy and Economics 87, 93–100.

14. Mori, S., Yamaji, K., Ishida, A., Prokushkin, S.G., Masyagina, O.V., Hagihara, A., Rafiqul Hoque, A.T.M., Suwa, A., Osawa, A., Nishizono, T., et al. (2010). Mixed-power scaling of whole-plant respiration from seedlings to giant trees. Proceedings of the National Academy of Sciences 107, 1447–1451.

15. Angelsen, A., Brockhaus, M., Sunderlin, W.D., and Verchot, L.V. (2012). Analysing REDD+: Challenges and choices (CIFOR).

16. Ojanen, P., and Minkkinen, K. (2020). Rewetting offers rapid climate benefits for tropical and agricultural peatlands but not for forestry-drained peatlands. Global Biogeochemical Cycles 34, e2019GB006503.

17. Unger, N. (2014). Human land-use-driven reduction of forest volatiles cools global climate. Nature Climate Change 4, 907–910.

18. Lawrence, D., Coe, M., Walker, W., Verchot, L., and Vandecar, K. (2022). The unseen effects of deforestation: biophysical effects on climate. Frontiers in Forests and Global Change 5, 756115.

19. van der Werf, G.R., Morton, D.C., DeFries, R.S., Olivier, J.G.J., Kasibhatla, P.S., Jackson, R.B., Collatz, G.J., and Randerson, J.T. (2009). CO_2 emissions from forest loss. Nature Geoscience 2, 737–738.

20. Gidden, M.J., Gasser, T., Grassi, G., Forsell, N., Janssens, I., Lamb, W.F., Minx, J., Nicholls, Z., Steinhauser, J., and Riahi, K. (2023). Aligning climate scenarios to emissions inventories shifts global benchmarks. Nature 624, 102–108.

21. Henrich, J., Heine, S.J., and Norenzayan, A. (2010). The weirdest people in the world? Behavioral and Brain Sciences 33, 61–83.

22. Planck, M. (2014). Scientific autobiography: and other papers (Open Road Media).

23. Roebroek, C.T., Duveiller, G., Seneviratne, S.I., Davin, E.L., and Cescatti, A. (2023). Releasing global forests from human management: How much more carbon could be stored? Science *380*, 749–753.

Chapter 12

1. FAO (2020). World food and agriculture: statistical yearbook 2020.
2. Voora, V., Larrea, C., and Bermudez, S. (2020). Global market report: bananas (International Institute for Sustainable Development).
3. Canadell, J., Jackson, R.B., Ehleringer, J.R., Mooney, H.A., Sala, O.E., and Schulze, E.D. (1996). Maximum rooting depth of vegetation types at the global scale. Oecologia *108*, 583–595.
4. Tallavaara, M., Eronen, J.T., and Luoto, M. (2018). Productivity, biodiversity, and pathogens influence the global hunter-gatherer population density. Proceedings of the National Academy of Sciences *115*, 1232–1237.
5. Junker, L. (2006). Population dynamics and urbanism in premodern island Southeast Asia. In Urbanism in the preindustrial world (University of Alabama Press).
6. Swamy, V., and Pinedo-Vasquez, M. (2014). Bushmeat harvest in tropical forests: knowledge base, gaps and research priorities (CIFOR).
7. Jost, L. (2007). Partitioning diversity into independent alpha and beta components. Ecology *88*, 2427–2439.
8. Berg, Å., Ehnström, B., Gustafsson, L., Hallingbäck, T., Jonsell, M., and Weslien, J. (1994). Threatened plant, animal, and fungus species in Swedish forests: distribution and habitat associations. Conservation Biology *8*, 718–731.
9. Jung, K., Kaiser, S., Böhm, S., Nieschulze, J., and Kalko, E.K.V. (2012). Moving in three dimensions: effects of structural complexity on occurrence and activity of insectivorous bats in managed forest stands. Journal of Applied Ecology *49*, 523–531.
10. Larjavaara, M., Davenport, T.R.B., Gangga, A., Holm, S., Kanninen, M., and Tien, N.D (2019). Payments for adding ecosystem carbon are mostly beneficial to biodiversity. Environmental Research Letters *14*, 054001.
11. Fang, H.L., Baret, F., Plummer, S., and Schaepman-Strub, G. (2019). An overview of global leaf area index (LAI): methods, products, validation, and applications. Reviews of Geophysics *57*, 739–799.
12. Ogden, F.L., Crouch, T.D., Stallard, R.F., and Hall, J.S. (2013). Effect of land cover and use on dry season river runoff, runoff efficiency, and peak storm runoff in the seasonal tropics of Central Panama. Water Resources Research *49*, 8443–8462.
13. Ellison, D., Morris, C.E., Locatelli, B., Sheil, D., Cohen, J., Murdiyarso, D., Gutierrez, V., van Noordwijk, M., Creed, I.F., Pokorny, J., et al. (2017). Trees, forests and water: cool insights for a hot world. Global Environmental Change-Human and Policy Dimensions *43*, 51–61.

14. Staal, A., Tuinenburg, O.A., Bosmans, J.H.C., Holmgren, M., van Nes, E.H., Scheffer, M., Zemp, D.C., and Dekker, S.C. (2018). Forest-rainfall cascades buffer against drought across the Amazon. Nature Climate Change *8*, 539–543.

15. Zheng, F.L., He, X.B., Gao, X.T., Zhang, C., and Tang, K.L. (2005). Effects of erosion patterns on nutrient loss following deforestation on the Loess Plateau of China. Agriculture Ecosystems & Environment *108*, 85–97.

16. Ghazoul, J. (2015). Forests: a very short introduction (Oxford University Press).

17. Watts, D., Matilainen, A., Kurki, S.P., Keskinarkaus, S., and Hunter, C. (2017). Hunting cultures and the 'northern periphery': exploring their relationship in Scotland and Finland. Journal of Rural Studies *54*, 255–265.

18. Holmes, T.P., Bowker, J.M., Englin, J., Hjerpe, E., Loomis, J.B., Phillips, S., and Richardson, R. (2016). A synthesis of the economic values of wilderness. Journal of Forestry *114*, 320–328.

19. Birge, T., and Herzon, I. (2014). Motivations and experiences in managing rare semi-natural biotopes: a case from Finland. Land Use Policy *41*, 128–137.

20. Stokland, J.N., Siitonen, J., and Jonsson, B.G. (2012). Biodiversity in dead wood (Cambridge University Press).

21. Soulé, M.E., and Wilcox, B.A. (1980). Conservation biology: an evolutionary-ecological perspective (Sinauer Associates).

Chapter 13

1. Popkin, G. (2019). How much can forests fight climate change? Nature *565*, 280–283.

2. Newby, A.G. (2023). Finland's great famine, 1856–68 (Springer).

3. DellaSala, D.A., Mackey, B., Norman, P., Campbell, C., Comer, P.J., Kormos, C.F., Keith, H., and Rogers, B. (2022). Mature and old-growth forests contribute to large-scale conservation targets in the conterminous United States. Frontiers in Forests and Global Change *5*, 979528.

4. Larjavaara, M. (2000). Kiertoajan suorat vaikutukset. Metsätieteen aikakauskirja *3*, 483–484.

5. de Jong, W., Liu, J., and Long, H. (2021). The forest restoration frontier. Ambio *50*, 2224–2237.

Chapter 14

1. Jung, M., Arnell, A., De Lamo, X., García-Rangel, S., Lewis, M., Mark, J., Merow, C., Miles, L., Ondo, I., Pironon, S., et al. (2021). Areas of global importance for conserving terrestrial biodiversity, carbon and water. Nature Ecology & Evolution *5*, 1499–1509.

2. Larjavaara, M., Chen, X., and Luo, M. (2024). A temperature-based model of biomass accumulation in humid forests of the world. Frontiers in Forests and Global Change *7*, 1142209.

3. Mittermeier, R.A., Turner, W.R., Larsen, F.W., Brooks, T.M., and Gascon, C. (2011). Global biodiversity conservation: the critical role of hotspots. In Biodiversity hotspots: Distribution and protection of conservation priority areas (Springer).

4. Hua, F., Bruijnzeel, L.A., Meli, P., Martin, P.A., Zhang, J., Nakagawa, S., Miao, X., Wang, W., McEvoy, C., Peña-Arancibia, J.L., et al. (2022). The biodiversity and ecosystem service contributions and trade-offs of forest restoration approaches. Science *376*, 839–844.

5. Shah, N.W., Baillie, B.R., Bishop, K., Ferraz, S., Högbom, L., and Nettles, J. (2022). The effects of forest management on water quality. Forest Ecology and Management *522*, 120397.

6. Brown, H.C.A., Appiah, M., and Berninger, F.A. (2022). Old timber plantations and secondary forests attain levels of plant diversity and structure similar to primary forests in the West African humid tropics. Forest Ecology and Management *518*, 120271.

7. Hua, F., Xu, J., and Wilcove, D.S. (2018). A new opportunity to recover native forests in China. Conservation Letters *11*, e12396.

8. Ray, D.K., Mueller, N.D., West, P.C., and Foley, J.A. (2013). Yield trends are insufficient to double global crop production by 2050. PloS One *8*, e66428.

9. Ramankutty, N., Mehrabi, Z., Waha, K., Jarvis, L., Kremen, C., Herrero, M., and Rieseberg, L.H. (2018). Trends in global agricultural land use: implications for environmental health and food security. Annual Review of Plant Biology *69*, 789–815.

10. Lal, R. (2010). Beyond Copenhagen: mitigating climate change and achieving food security through soil carbon sequestration. Food Security *2*, 169–177.

11. Ritzema, H., Limin, S., Kusin, K., Jauhiainen, J., and Wosten, H. (2014). Canal blocking strategies for hydrological restoration of degraded tropical peatlands in Central Kalimantan, Indonesia. Catena *114*, 11–20.

12. Siarudin, M., Rahman, S.A., Artati, Y., Indrajaya, Y., Narulita, S., Ardha, M.J., and Larjavaara, M. (2021). Carbon sequestration potential of agroforestry systems in degraded landscapes in West Java, Indonesia. Forests *12*, 714.

13. Pardon, P., Reubens, B., Mertens, J., Verheyen, K., De Frenne, P., De Smet, G., Van Waes, C., and Rehcul, D. (2018). Effects of temperate agroforestry on yield and quality of different arable intercrops. Agricultural Systems *166*, 135–151.

14. Tesfahun, B., Kebede, K., and Effa, K. (2017). Traditional goat husbandry practice under pastoral systems in South Omo zone, southern Ethiopia. Tropical Animal Health and Production *49*, 625–632.

15. Hayek, M.N., Piipponen, J., Kummu, M., Resare Sahlin, K., McClelland, S.C., and Carlson, K. (2024). Opportunities for carbon sequestration from removing or intensifying pasture-based beef production. Proceedings of the National Academy of Sciences *121*, e2405758121.

Chapter 15

1. Masson-Delmotte, V., Zhai, P., Pirani, A., Connors, S.L., Péan, C., Berger, S., Caud, N., Chen, Y., Goldfarb, L., Gomis, M., et al. (2021). Climate change 2021: the physical science basis. Contribution of Working Group I to the Sixth Assessment Report of the Intergovernmental Panel on Climate Change (IPCC).

2. Wei, S., Yi, C., Fang, W., and Hendrey, G. (2017). A global study of GPP focusing on light-use efficiency in a random forest regression model. Ecosphere 8, e01724.

3. Larjavaara, M., Lu, X., Chen, X., and Vastaranta, M. (2021). Impact of rising temperatures on the biomass of humid old-growth forests of the world. Carbon Balance and Management 16, 31.

4. George, S.S. (2014). An overview of tree-ring width records across the Northern Hemisphere. Quaternary Science Reviews 95, 132–150.

5. Brienen, R.J., Schöngart, J., and Zuidema, P.A. (2016). Tree rings in the tropics: Insights into the ecology and climate sensitivity of tropical trees. In Tropical tree physiology (Springer).

6. Berg, A., and Sheffield, J. (2018). Climate change and drought: the soil moisture perspective. Current Climate Change Reports 4, 180–191.

7. Norby, R.J., and Zak, D.R. (2011). Ecological lessons from free-air CO_2 enrichment (FACE) experiments. Annual Review of Ecology, Evolution, and Systematics 42, 181–203.

8. Du, E., Fenn, M.E., De Vries, W., and Ok, Y.S. (2019). Atmospheric nitrogen deposition to global forests: Status, impacts and management options Environmental Pollution, 250, 1044–1048.

9. Himes, A., Bauhus, J., Adhikari, S., Barik, S.K., Brown, H., Brunner, A., Burton, P.J., Coll, L., D'Amato, A.W., Diaci, J., et al. (2023). Forestry in the face of global change: results of a global survey of professionals. Current Forestry Reports 9, 473–489.

10. Henttonen, H.M., Nöjd, P., and Mäkinen, H. (2024). Environment-induced growth changes in forests of Finland revisited-a follow-up using an extended data set from the 1960s to the 2020s. Forest Ecology and Management 551, 121515.

11. Wright, S.J. (2010). The future of tropical forests. Annals of the New York Academy of Sciences 1195, 1–27.

12. Jones, M.W., Abatzoglou, J.T., Veraverbeke, S., Andela, N., Lasslop, G., Forkel, M., Smith, A.J., Burton, C., Betts, R.A., van der Werf, G.R., et al. (2022). Global and regional trends and drivers of fire under climate change. Reviews of Geophysics 60, e2020RG000726.

13. Kahneman, D. (2011). Thinking, fast and slow (Macmillan).

Chapter 16

1. Cheng, Z., Aakala, T., Ji, C., and Larjavaara, M. (2025). Disturbance dynamics and its effects on carbon in human-impacted mountain forests in northwestern Yunnan, China. Ecology and Evolution 15, e72165.

2. Wullenkord, M.C., and Reese, G. (2021). Avoidance, rationalization, and denial: defensive self-protection in the face of climate change negatively predicts pro-environmental behavior. Journal of Environmental Psychology 77, 101683.

3. Sanders, A.J., Ford, R.M., Mulyani, L., Larson, A.M., Jagau, Y., and Keenan, R.J. (2019). Unrelenting games: multiple negotiations and landscape transformations in

the tropical peatlands of Central Kalimantan, Indonesia. World Development *117*, 196–210.

4. Henrich, J. (2020). The WEIRDest people in the world: how the west became psychologically peculiar and particularly prosperous (Farrar, Straus and Giroux).

5. Pulkkinen, K., Undorf, S., Bender, F., Wikman-Svahn, P., Doblas-Reyes, F., Flynn, C., Hegerl, G.C., Jönsson, A., Leung, G.-K., and Roussos, J. (2022). The value of values in climate science. Nature Climate Change *12*, 4–6.

6. Shrout, P.E., and Rodgers, J.L. (2018). Psychology, science, and knowledge construction: Broadening perspectives from the replication crisis. Annual Review of Psychology *69*, 487–510.

www.ingramcontent.com/pod-product-compliance
Ingram Content Group UK Ltd.
Pitfield, Milton Keynes, MK11 3LW, UK
UKHW030740010526
470603UK00011B/192